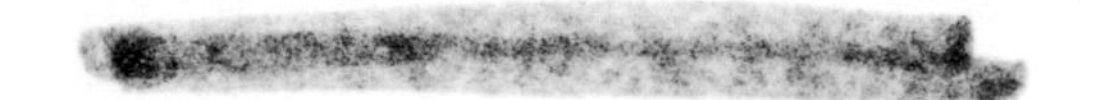

Reliability of Transport Networks

TRAFFIC ENGINEERING SERIES

Series Editor: **Professor Michael G H Bell**
University of Newcastle, UK

3. Capturing Long Distance Travel
Edited by K Axhausen, J-L Madre, J Polak *and* P Toint *

4. Reliability of Transport Networks
Edited by Michael G H Bell *and* Chris Cassir

* *forthcoming*

Reliability of Transport Networks

Edited by

Michael G H Bell *and* Chris Cassir
Department of Civil Engineering, University of Newcastle, UK

RESEARCH STUDIES PRESS LTD.
Baldock, Hertfordshire, England

RESEARCH STUDIES PRESS LTD.
16 Coach House Cloisters, 10 Hitchin Street, Baldock, Hertfordshire, England, SG7 6AE
and
325 Chestnut Street, Philadelphia, PA 19106, USA

Marketing:

UK, EUROPE & REST OF THE WORLD
Research Studies Press Ltd.
16 Coach House Cloisters, 10 Hitchin Street, Baldock, Hertfordshire, England, SG7 6AE

Distribution:

NORTH AMERICA
Taylor & Francis Inc.
International Thompson Publishing, Discovery Distribution Center, Receiving Dept., 2360 Progress Drive Hebron, Ky. 41048

ASIA-PACIFIC
Hemisphere Publication Services
Golden Wheel Building # 04-03, 41 Kallang Pudding Road, Singapore 349316

UK, EUROPE & REST OF THE WORLD
John Wiley & Sons Ltd.
Shripney Road, Bognor Regis, West Sussex, England, PO22 9SA

Library of Congress Cataloging-in-Publication Data

Reliability of transport networks / edited by Michael G. H. Bell and Chris Cassir.
p. cm. -- (Traffic engineering series ; 3)
Includes bibliographical references and index.
ISBN 0-86380-260-5
1. Traffic engineering. 2. Choice of transportation. 3. Transportation--Planning. I. Bell, Michael G. H. II. Cassir, Chris, 1970-III. Series.

HE333 .R44 2000
338.3--dc21 99-088744

British Library Cataloguing in Publication Data
A catalogue record for this book is available from the British Library.

ISBN 0 86380 260 5

Printed in Great Britain by SRP Ltd., Exeter

Contents

INTRODUCTION

Chris Cassir[1], *Michael Bell*[1] *and Yasunori Iida*[2]

IMPORTANCE OF TRANSPORT NETWORKS

The efficient functioning of any society depends critically on networks of various kinds, such as water supply, energy supply, sewage disposal, communication and, of course, transportation. The importance of transportation is perhaps best appreciated in situations where it is severely disrupted, for example by an earthquake, as occurred in Kobe, Japan, on 17 January 1995.

The potential sources of disruption to transportation networks are numerous, ranging from natural or man-made disasters (like earthquakes, floods, landslides, terrorist attacks, mining subsidence, bridge or tunnel collapses, and major accidents) at one extreme, to routine events (like congestion, road maintenance, badly-parked vehicles, and minor collisions) at the other. The scale, impact, frequency and predictability of such events will of course vary enormously. While little can be done about their scale, frequency or predictability, particularly where natural disasters are concerned, it is possible to design and manage transportation networks so as to minimise the disruption that such events can cause.

RELIABILITY IN PLANNING AND OPERATION

Despite being fundamental for good planning, uncertainty is not well accounted for in many tools for analysis and forecasting that are routinely used to provide quantitative input to the design and management of transport networks. Such tools include traffic assignment models, like the CONTRAM or SATURN traffic simulators, which provide predictions of average flows and travel times in a road network. Such average values are, however, not sufficient for the assessment of transport related schemes, since they do not reflect the extent of variability, or show the robustness of the network in coping with such variability. Probabilistic

[1] Transport Operations Research Group (TORG), University of Newcastle (M.G.H.Bell@ncl.ac.uk)
[2] Department of Transport Engineering, Kyoto University (iida@urbanfac.kuciv.kyoto-u.ac.jp)

performance measures are required, and *network reliability* is one such set of measures. Reliability measures, used extensively for engineering systems in order to assess the stability of the quality of service offered, ought to be given more consideration in the planning and operation of transport networks than is currently the case, especially since uncertainty of one kind or another is an important characteristic of transport systems.

It is surprising that reliability analysis has been lacking in the appraisal of transport networks. A possible explanation for this may relate to the difficulty in defining and measuring reliability, particularly for transport networks. In systems engineering, reliability is generally calculated as the probability that the system can perform its desired function to an acceptable level of performance, for some given period of time. However, in the case of transport networks, neither the desired function nor the corresponding acceptable level of performance lend themselves to straightforward and unambiguous definition.

There are, in fact, many functions or objectives that may be associated with a transport network, in addition to its most basic one, namely the physical transport of people and goods from one location to another (the connectivity function). Depending on the viewpoint of the actors concerned (network planners, network managers, network users, etc.), the function of a transport network can be described in terms of travel time, throughput, accessibility, equity, etc. Different functions lead to different performance measures. Still, provided the reliability of a transport network is appropriately defined with respect to some function, there is a basis for defining a useful indicator of network performance which considers the uncertainty associated with transportation systems and user demands.

RELIABILITY MEASURES

Reliability, however defined, provides a measure of the stability of the quality of service, which the transport system can offer to its users. This should have important implications for the planning, construction and management of transport networks. There are two ways to achieve reliability by design. One is to *over design* network components. For example, a bridge could be made stronger than it would normally need to be. The other is to build in a degree of *redundancy*. For example, two river crossings could be built where one would normally suffice. Having produced a design, network reliability can be assessed in terms the probability that a trip can be completed to a specified level of network performance (measured in terms of travel time, cost, etc.).

Another application of reliability analysis could be to traffic management. With the advent of Intelligent Transport Systems (ITS), it is possible to design network management systems that ensure smooth travel under normal traffic fluctuations and also limit the impacts of unexpected disruptions, like accidents. ITS such as Dynamic Route Guidance (DRG) or Variable Message

Signs (VMS) are generally aimed at providing users with a high level of average network performance, typically measured in terms of travel times. But given the uncertainty attached to actual traffic conditions and the difficulties associated with their prediction, these schemes should seek to improve network reliability, which could be measured by the probability that a trip can be completed within a given time.

DECISION-MAKING UNDER UNCERTAINTY

A major factor in the determination and improvement of reliability is a good understanding and representation of the complex interactions between travel behaviour and network performance, in particular the response of travellers to uncertainty or adverse situations. The presence of network users reacting autonomously to network conditions makes transport network reliability different from other types of system reliability. One of the major difficulties in the analysis lies in the choice of appropriate routing models to represent travellers' movements in an uncertain environment, which should include the feedback effect of those movements on uncertainty.

Although some assignment models include a degree of uncertainty in certain variables, like perceived travel times in Stochastic User Equilibrium (SUE) assignment, their outputs tend to resemble deterministic values. The feedback from the decisions taken by network users to the factors behind their decision-making should include the effects of random variation. Haselton (1997) considered the case where travel time, assumed to be the basis of route choice, is itself random due to randomness in decision-making, adding an extra component to random variation.

If travellers routinely misperceive actual travel times, as assumed in SUE assignment, there comes a point where they would become aware of this and start to modify their behaviour accordingly, perhaps by building safety margins into their selection processes. For example, if route k is known to be more unreliable than alternative route k', then in making a choice a larger safety margin will be given to route k than route k'. Uchida and Iida (1990) present a useful expression for the optimal safety margins in route choice behaviour.

While safety margins are without doubt a part of every day decision-making, the notion that travellers are in a position to optimise them makes rather strong assumptions about the amount and quality of information available to them. It presumes that travellers have accurate knowledge of the size of travel time variation, assuming for the moment that travel time is the sole basis for a choice. But where would travellers obtain such information? Perhaps it is learnt from day-to-day experience. Over a long period of time, travellers could in theory progressively improve their safety margins. While this scenario may describe the daily journey to work, there are many situations where travellers are not able to acquire sufficient experience of travel time variability in order to optimise their safety margins. How would the rational traveller behave then?

Consider a traveller with a choice of routes but no idea of the travel time variability on each. The following might be the sort of mental process he would adopt. He would start by picking the best route if nothing goes wrong, consider what might go wrong, and then respond to this by picking another route that would avoid the potential difficulty, consider what might go wrong on this route, pick another route, etc. Suppose route k is best if nothing goes wrong, but if link i is blocked, then route k' would be a good alternative, unless link i' were blocked, in which case route k'' would be a good alternative, etc. He doesn't know which (if any) link is blocked, but thinks that the likelihood of two links being blocked is negligible. The process just described corresponds to a two-person, non-cooperative, zero sum game. The mixed strategy Nash equilibrium gives the probability that the traveller thinks any route is best and the corresponding fears that any link is blocked. Chapter 7 goes into greater detail.

CONTENTS OF THE BOOK

This book, based on the papers presented at a Workshop on Transport Network Reliability held at the University of Newcastle in August 1999, presents recent developments in the field of transport network reliability. The general aim of the book is to shed some light on the need for, the evaluation of, and potential use of reliability, as a performance indicator for transportation networks. The chapters can be grouped into three sets, reflecting different issues concerning the study of transport network reliability as a whole.

The first set is concerned with traffic uncertainty and travellers behaviour under uncertainty. The first chapter presents some interesting results of a survey experiment showing that, while there seems to be a general aversion of travellers to uncertainty, some people appeared to respond to higher levels of uncertainty by choosing riskier options. This suggests that travellers are affected by potential gains that would be foregone if the risk is not accepted. The second chapter introduces a model of day-to-day learning of expected travel times under uncertainty, and fits the model to some empirical data. The third chapter proposes an extension to the traditional concept of a Stochastic User Equilibrium assignment, by endogeneously calculating the moments of traffic flow and travel time probability distributions. The fourth and fifth chapters both consider the inherent uncertainty and instability associated with traffic conditions and their attempted control, while the sixth chapter presents a new routing and scheduling model for pick-up/delivery trucks that takes variable travel times explicitly into account.

The second set of chapters suggest various methods for the assessment of reliability in transport networks. The seventh chapter uses game theory concepts to derive some measures of reliability, while taking users responses and their feedback on expected travel times into account. The eighth chapter proposes a method for the assessment of motorway reliability in the face of events such as disasters, congestion and bad weather. The ninth chapter balances the definitions of travel time reliability versus capacity reliability in a network, and shows how they are inter-related. Finally, the tenth chapter presents a network traffic flow simulator that can be used to estimate travel time probability distributions for the evaluation of travel time reliability, and presents a Genetic Algorithm method for calibrating the parameters in the simulator.

The last set of chapters hints at some possible applications of reliability in network planning and traffic management. The eleventh chapter looks at how to evaluate the maximum flow that a network can carry under various link travel time and link capacity reliability conditions. The twelfth chapter proposes a new equilibrium assignment model, similar to Deterministic User Equilibrium, but based on the maximisation of route reliability, where reliability is defined as the probability of not exceeding the capacity of a link (congestion). Following the same general idea, the thirteenth chapter introduces a static and dynamic routing model that tries to maximise travel time reliability. The fourteenth chapter also introduces a routing model, which reflects driver behaviour in degraded networks. It is shown how in this case the provision of information can increase the expected value of network performance. Finally, the last chapter suggests the application of some reliability indices related to connectivity and travel times in order to assess traffic calming schemes.

REFERENCES

Haselton, M.L (1997) Some Remarks on Stochastic User Equilibrium *Transportation Research B,* Vol 32, No 2, pp 101-108.
Uchida, T and Iida, I (1993) Risk assignment: A new traffic assignment model considering risk of travel time variation. *Proc. 12th International Symposium on transportation and Traffic Theory* (Ed. C F Daganzo), Elsevier, 89-105.

CHAPTER 1

TRAVELLERS' RESPONSE TO UNCERTAINTY

Peter Bonsall, Institute for Transport Studies, University of Leeds, Leeds, LS2 9JT
pbonsall@its.leeds.ac.uk

1 INTRODUCTION

This Chapter begins with a discussion of uncertainty as a psychological, rather than as a purely statistical, concept. This discussion will provide the background for a review of the strategies adopted by different individuals when faced with uncertainty. Some evidence will be presented which suggests that, contrary to the usual assumption that uncertainty is an undesirable characteristic in transport systems, some people seem to derive benefit from it – a result which raises interesting questions about the valuation of uncertainty.

1.1 The Nature of Uncertainty

Uncertainty, unpredictability and unreliability are related but subtly different concepts. Although they *may* each result from the variability which affects most systems, variability does not inevitably lead to uncertainty, unpredictability or unreliability. If the variability contains a systematic element it may, to some extent, be predictable and thus, to that extent, neither uncertain nor unreliable. However, even if something is *theoretically* predictable, if it cannot be predicted in practice then it is *effectively* unpredictable.

Reliability implies a degree of predictability but not necessarily complete invariance or stability, indeed something may be regarded as "reliably variable". Authors elsewhere in this book define unreliability as an unacceptable level of uncertainty thus implying a degree of unpredictability (but, as indicted above, not necessarily of variability). Other authors ignore the question of knowledge and suggest that reliability is directly related to variability and may simply be measured via the probability distribution (eg the probability of arriving at a destination within a predetermined period).

I argue that, to understand responses to uncertainty, we must recognise that it is state of mind rather than simply a statistical concept. If travellers are unaware that something is, in the statistical sense, certain, they will respond as if it were uncertain. The reverse is also true; if travellers are unaware that something is, in the statistical sense, uncertain, they will respond as if it were certain. (A perverse consequence of this is that the provision of information about the magnitude of uncertainty can lead travellers to respond as if the uncertainty had increased even though it has in fact decreased.)

1.2 Attitudes to Risk

Expected Utility Theory (see Von Neumann and Morgenstern, 1944) has dominated the literature on behavioural response to uncertainty. The theory suggests that behaviour can be explained as the result of decision makers choosing those actions which maximise the expected utility (EU) of the alternative courses of action. The expected utility of each action will be the sum of the utilities of all the potential outcomes of that course of action multiplied by their respective probabilities – as summarised in equation (1)

$$EU_a = \Sigma \, (U_{oa} \cdot P_{oa}) \qquad (1)$$

Where EU_a is the expected utility of course of action a,

U_{oa} is the utility of outcome o of action a, and

P_{oa} is the expected probability of outcome o of action a.

Thus, if $P_{1a}=0.5,\ P_{2a}=0.3,\ P3_{2a}=0.2,\ U_{1a}=10,\ U_{2a}=20$ and $U3_a= 40,$ then $EU_a=5+6+8=19$

In order to reflect reality, the calculation of utilities (U_{oa}) must of course allow for context-specificity, non-linearity and asymmetric valuation. The point may be illustrated with reference to the following examples:

(1) the consequences of a given delay may be more serious on a time-constrained journey than on one that is not so constrained – thus the consequences of delays on work-related journeys will generally be more serious than those of delays on leisure journeys and the consequences of a delay on a road leading *towards* an airport will generally be more serious than those of a delay on a road leading *away from* an airport.

(2) the consequences of a ten minute delay may be more than twice as severe as those of a five minute delay

(3) the consequences of an extra five minute delay during a journey may depend on how much delay has so far accrued on that journey

(4) the utility of an unexpected five minute decrease in journey time may be less than the disutility of an unexpected five minute increase in journey time.

Expected Utility Theory can cope with such complications (given the necessary data) but requires a strictly consistent application of the expected probabilities of the various outcomes. Unfortunately, however, many people are far from consistent in their assessment of probabilities. Some departures from a strictly consistent assessment of probabilities are well documented; it is, for example, well established that most people exaggerate low probabilities (*vide* the popularity of lotteries and the fear of flying) and ignore differences between low probabilities (for example, odds of 1:2,000,000 and 1:3,000,000 would generally be treated as virtually identical). Many departures from a strictly consistent attitude to probabilities are associated with personality traits; optimists like to gamble – they are risk seeking, while pessimists do not like risking loss (or failure to win) - they are risk averse.

Kahneman and Tversky's Prospect Theory (1979) was developed to explain commonly observed departures from the behaviour that would be predicted by Expected Utility Theory. Its key tenet is that decisions are crucially context dependent and that the evaluation of risky prospects involves a sequential assessment of outcomes during which process the prospects are disassembled, simplified and reassembled in ways that can result in inconsistent preferences. Kahneman and Tversky demonstrated how such processes could explain apparently irrational responses to uncertainty and that, although we should expect risk averse behaviour when the outcomes involve gains, we should expect risk seeking behaviour when the outcomes involve losses.

1.3 Travellers' Potential Strategies for Dealing with Uncertainty

Strategies which travellers could adopt to cope with uncertainty (in travel time) will include:

1 Seeking to minimise variability:

1.1 by choosing routes on which the conditions are relatively stable,

1.2 by travelling at times when conditions are relatively stable, or

1.3 by varying speed inversely with the ambient conditions.

2. Making maximum use of existing knowledge:

2.1 by sticking to well-known routes, or

2.2 by travelling at their "usual" times of day.

3. Seeking to increase their knowledge:

3.1 by experimenting to build up knowledge, or

3.2 by making use of all available information sources.

4. Seeking to minimise consequences of unreliability:

4.1 by building safety margins into their schedule, or

4.2 by phoning ahead to inform people at the destination of their likely arrival time.

5. Capitalising on the uncertainty by enjoying a game of chance:

5.1 against the system,

5.2 against other drivers, or

5.3 against themselves-on-other-days.

Examples of these strategies will be familiar to all drivers but, with notable exceptions such as the work on schedule safety margins (relevant to strategy *4.1*) and more recently on information acquisition strategies (relevant to strategy *3.2*), most of them have been ignored in models of traveller behaviour. Lee, Moon and Asakura (in Chapter 12) and Maher and Zhang (in Chapter 13), address the proposition that route choices might reflect a desire to minimise uncertainty (relevant to strategies *1.1* and *1.2*) but work on the other strategies is rare.

A major problem has been the shortage of evidence with which to back up the theories or calibrate the models. Notable exceptions to this general rule again include work on trip scheduling (eg the classic work by Small, 1982) and on information acquisition (see, for example, Polak and Jones, 1993 or Walker and Ben-Akiva, 1994). More recent evidence, particularly relevant to strategy 2.1, has been provided by stated preference surveys which have shown that, when under time pressure, drivers are more reluctant to accept advice to depart from familiar routes (Bonsall et al 1995) and by the VLADIMIR route choice simulator which has revealed clear evidence of the way in which drivers making unfamiliar journeys seek to make maximum use of routes with which they are already familiar (Bonsall et al, 1997).

Numerous studies in experimental economics have emphasised the role of gambling, or risk-seeking behaviour. Recent examples relating to travel behaviour, and particularly to strategy 5.1, are provided by Powell and Davis (1996) and Delvert and Petiot (1998) who have used experimental economics to infer attitudes to risk among drivers. Powell and Davis concluded that interurban route choice strategies could be interpreted as "games" played by drivers seeking to outperform the expected duration for the journey in question. Delvert and Petiot found some evidence to suggest that drivers would generally prefer to risk an uncertain journey duration rather than pay a toll to secure a certain journey duration. The strategies adopted by participants in games of chance constructed by experimental economists may or may not reflect real-life behaviour; it can certainly be argued (see Bonsall, 1997) that such experiments encourage game-playing and hence any interpretation of the data as evidence of real-life risk seeking is bound to be fraught. It would obviously be preferable to find a data source which is not so open to the possibility of having provided an exaggerated incentive to gamble.

2 A NEW SOURCE OF DATA

2.1 Cho's Study

A useful data set was collected by Cho as part of her PhD studies at Leeds (1998) and has been re-analysed by Bonsall and Cho in a recent paper (1999). The data source was a stated preference questionnaire designed to explore drivers' responses to imprecisely-defined road tolls. The questionnaire, to which 160 drivers responded, offered choices between routes with road tolls defined with different levels of precision and revealed intriguing evidence on responses to uncertainty. The results are shown in Table 1. (Note that there were three groups of respondents; each having a different toll on route1 - £1.00, £1.50 and £2.00, respectively.)

The data in Table 1 show a general preference for routes with precisely-known tolls over routes with imprecisely-known tolls. This is indicative of a general aversion to risk and conforms with conventional expectation. However, the data also suggest that, for all three groups, the preference for the precisely-known charge was *lowest* when the uncertainty was greatest. In other words, the tendency to avoid the uncertain prospect is reduced when the uncertainty is most marked. In an attempt to throw light on this phenomemon, the data are disaggregated in Table 2 according to the respondent's gender and stated household income. (Note that the data in Table 2 are aggregated over the three groups identified in Table 1.)

Table 1: Choice between routes with tolls known with different levels of precision.

Group	Toll on route1 (£)	Toll on route2 (£)	% choosing route1
A	1.00	0.90 – 1.10	*76*
	1.00	0.80 – 1.20	*76*
	1.00	0.70 – 1.30	*71*
B	1.50	1.35 – 1.65	*65*
	1.50	1.20 – 1.80	*68*
	1.50	1.05 – 1.95	*61*
C	2.00	1.80 – 2.20	*72*
	2.00	1.60 – 2.40	*72*
	2.00	1.40 – 2.60	*59*

Table 2: Different groups' preference for the route with the most precisely known toll.

Toll on route2 relative to that on route1	% choosing route1					
		gender		household income (£k/yr)		
	all	male	female	<20	20-40	>40
route1 ± 10%	*72*	*69*	*76*	*94*	*71*	*66*
route1 ± 20%	*72*	*66*	*80*	*61*	*74*	*74*
route1 ±30%	*63*	*57*	*71*	*50*	*62*	*70*

The data in Table 2 suggest that the females are more risk-averse than the males and that the males become steadily less risk-averse as the level of uncertainty increases. All groups except the highest income group are least risk-averse when the uncertainty is ±30%. The lowest income group is very risk-averse at relatively low levels of uncertainty (±10%) but appears less risk-averse when the uncertainty is highest (±30%). It seems that the group for whom the loss would be most serious is the most ready to accept the risk.

2.2 Interpretation of Cho's Results

These results appear inconsistent with strict application of Expected Utility Theory. There are, however, at least four interpretations of the data shown in Tables 1 and 2. They are, respectively, that the result is an artifact of Cho's experimental design, that it is consistent with Prospect Theory, that it results from the respondent's assumptions about the shape of the probability distribution and that it is a manifestation of gambling behaviour. I will deal with each of these in turn.

The first possibility that one should always consider when faced with unexpected experimental results (or expected ones for that matter!) is that they are an artifact of the experimental process itself. The possibility existed in the current case that, since the questions whose results are presented in Table 1 were ordered such that the degree of uncertainty about route 2's toll increased from one question to the next, the apparent reduced aversion to uncertainty might simply have been the result of a regression to the mean brought on by respondent fatigue. In fact this explanation is not supported by closer examination. The series of three questions whose results are presented in Table 1 were each part of a sequence of seven and the reducing preference for the certain toll was not apparent throughout these sequences – even though the final questions in the sequence were more complicated and so should have been most subject to any fatigue effect.

Cho (1998) sought to explain the results using Prospect Theory but found the theory unsatisfactory as an explanation of why a general preference for the certain prospect should co-exist with an increasing preference for the increasingly uncertain prospect.

The third interpretation of the data in Tables 1 and 2 is that the respondents inferred a distribution to the probabilities within the range of uncertainty presented to them, such that it became quite 'rational' to behave as they did. The respondents were told that the charge would lie somewhere in the given range. They were not told anything about the distribution of probabilities within this range. However, they were told that the reason for the uncertainty was that the charges were a function of travel times. If they thought about this they may have concluded that the distribution of probabilities within the given range was not uniform but rather was skewed to the left (towards lower travel times and charges) with a tail extending out to higher travel times and charges – as summarised in Figure 1. If this were the case then it might be logical for them to assume that, as the range of uncertainty increased, so too would the extent of the skew (compare Figures 1a and 1b). Since the median of a left-skewed distribution would be to the left of the midpoint of the range, it would be completely logical to prefer the uncertain charge chosen from the skewed distribution to a certain charge equal to the midpoint of the range. The reason for the reduction in the initial preference for the certain toll would be that, although an aversion to risk remains, some people take account of the fact that the midpoint accelerates rightward more rapidly than does the median. This explanation of Cho's results has the attraction of being consistent with Expected Utility Theory but requires us to accept that some respondents (particularly males and people with low household incomes) were imagining increasingly skewed distributions of tolls and thinking through the logic outlined above.

Figure 1: Effect of increasing skew as range of uncertainty increases.

(a) medium range of uncertainty **(b) considerable range of uncertainty**

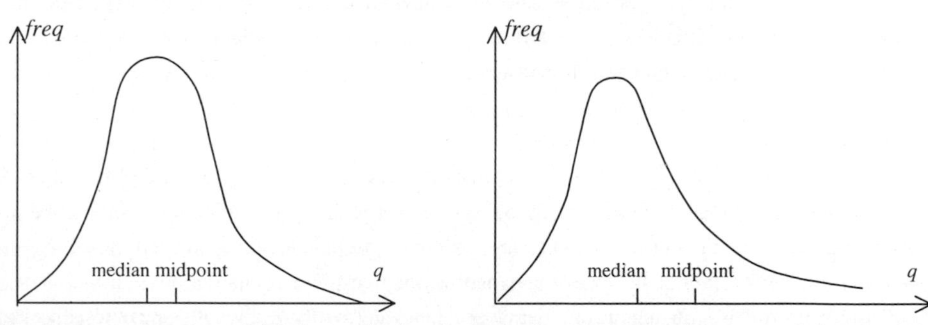

The fourth explanation of Cho's results has the attraction of not requiring us to infer that the respondents engaged in such mental gymnastics. It is that, when faced with an uncertain prospect, some people will adopt a gambler's strategy; they will deliberately choose the option which offers the greatest potential gain even though it brings the risk of a possible loss. They may do this because they have a naturally optimistic outlook on the world, or perhaps because they have high confidence in their own judgement or wish to retain control of the situation, or perhaps simply because they get a buzz from confronting the risk. The data in Tables 1 and 2 suggest that this gambling response is most pronounced among males and when the risk is non-trivial for the decision maker.

3 IMPLICATIONS

We have observed a general aversion to uncertainty (particularly among females and low income groups) but have noted that when the uncertainty becomes non-trivial some people switch back to the more uncertain prospect. These findings have important implications for the forecasting of individual response to uncertain prospects and, arguably, for the evaluation of reliability.

If the phenomenon is the result of sophisticated assumptions by some respondents about the shape of underlying probability distributions, then we must clearly take this into account in the design of survey instruments and in the analysis of data derived from them. We should take steps to ascertain what distribution is being assumed, by whom and why. The widespread assumption that respondents will, unless carefully instructed otherwise, assume a uniform or normal distribution would seem to be in doubt.

If the phenomenon reflects a tendency by at least some travellers deliberately to select uncertain prospects (whether from a belief that they can beat the odds, a desire to retain the locus of control or simply in order to enjoy the thrill of the gamble), it would obviously be over simplistic to assume, as do Lee, Moon and Asakura (in Chapter 12), that all drivers are seeking to minimise unreliability. Even if risk-seeking behaviour is restricted to a minority of the total population of drivers, it is likely to be worth disaggregating the population by its attitude to risk because interaction between the different sub-populations will probably affect the overall result.

Notwithstanding the above, it obviously remains desirable to reduce the waste of resources caused by unreliability. This Chapter should not be taken as a denigration of the role of traffic engineering and control, of enforcement (eg of on-street parking restrictions) or of the provision of information in reducing unreliability and its consequences. It does suggest, however, that network managers should be aware that people may respond to increased reliability, or to increased information, in unexpected ways and that they will not always appear to appreciate increased certainty or information as, rationally, they ought. This is partly because behaviour will be adjusted in the light of changes in expected reliability (with the consequence that, in a generally reliable system, the effect of the occasional breakdown will be felt even more strongly). But it is partly because there are some sectors of the driving population who seem to derive a perverse benefit from uncertainty. This leads to the heretical idea that such benefits might sometimes be given a positive value in appraisals and prompts the question of whether any equity issues are raised by the finding that different people have different attitudes to risk? (Would it be unfair on males to deprive them of the uncertainty in network conditions on which some of them seem to thrive?)

As ever, what we really need is more research to illuminate these issues. Carefully designed experiments might seek to throw light on the type of gambling strategies being adopted (are they, for example, consistent with what Game Theory would predict in a multi-actor system?) and on the value which appears to be being put on the opportunity to gamble. In designing these experiments, however, we must be careful not to get so divorced from real situations that the results relate only to the experimental paradigm being used.

REFERENCES

Bonsall, P.W. (1997). Motivating the respondent; how far should you go? Paper presented at IATBR'97 , Austin, Texas.

Bonsall, P.W. and Cho, H-J. (1999). Travellers response to Uncertainty, the particular case of drivers' response to imprecisely known tolls and charges. *Proceedings of European Transport Conference*, University of Cambridge, Sept 1999, PTRC, London

Bonsall, P.W., Whelan G.A. and Page M. (1995). Stated preference experiments to determine the impact of information on drivers route choice. *Proc 23rd European Transport Forum,* Seminar E, pp287-306, PTRC London 1995.

Bonsall, P.W., Firmin, P.E., Anderson, M. E., Palmer, I.A. and Balmforth, P.J. (1997). Validating the results of route choice simulator. *Transportation Research C*. 5(6). 1997.

Cho, H-J. (1998). Route choice responses to variable road user charges and traffic information, PhD Thesis, Institute for Transport Studies, University of Leeds.

Delvert, K. and Petiot R. (1998). Behavioural response to traffic variability: an experimental outlook on road pricing as willingness to pay for certainty. Paper presented at World Conference on transport research, Antwerp, July, 1998.

Kahneman, D. and Tversky A., (1979). Prospect Theory: an analysis of Decision under risk. *Econometrica* 47 no2 pp263-291.

Polak, J. and Jones, P.M. (1993). The acquisition of pre-trip information: a stated preference approach. *Transportation 20.3*

Powell, M. and Davis, A., (1996). Aggregating small time savings in UK trunk road scheme appraisals, Leeds University Business School Discussion Note E96/02.

Small, K. (1982). The scheduling of consumer activities. *American Econ. Review, 72,* 467-479.

Walker, J. and Ben-Akiva, M.E. (1994) Modelling Traveller response to traveller information systems: laboratory simulation of information searches using multimedia technology. Paper presented at 75th TRB Annual Meeting, Washington DC.

Wardman, M.R., Bonsall, P. W. and Shires, J. D. (1997). Stated preference analysis of driver route choice reaction to variable message sign information. *Transportation Research C*. Vol 5C no.6, 1997.

Chapter 2

An Empirical Model of Travellers' Day-To-Day Learning in the Presence of Uncertain Travel Times

John Polak and Frederic Oladeinde
Centre for Transport Studies
Department of Civil and Environmental Engineering
Imperial College of Science and Technology and Medicine
London SW7 2BU

Introduction

Understanding the dynamics of transport systems is widely regarded as one of the most important challenges facing travel behaviour researchers. One of the key behavioural processes underlying the dynamical properties of transport systems is the gradual evolution of travellers' expectations of travel conditions as a result of repeated day-to-day experience. This process of "adaptation through experience" can usefully be conceived of as a process of learning in an uncertain environment (Oladeinde, 2000). A number of recent simulation studies have highlighted the fact that the nature of this learning process can have a profound effect upon the temporal evolution and final equilibrium states of transport systems and upon the impact of policy interventions such as traveller information systems (e.g., Axhausen *et al.*, 1995; Emmerink *et al.*, 1996; Hazelton and Polak, 1997; Horowitz, 1984; Koutsopoulos, and Xu, 1993; Polak and Hazelton, 1998). However, despite the evident importance of these learning mechanisms, there has so far been surprisingly little work carried out to establish the empirical properties of travellers' learning mechanisms.

This chapter reports the results of a study which has developed and empirically estimated a simple theoretical framework for modelling travel time learning processes. This approach builds on and extends existing treatments of traveller learning in the literature. The chapter comprises several sections. In the first section we briefly review the existing literature on traveller learning and identify a number of important weaknesses in these approaches. The second section outlines a general behavioural framework for characterising traveller learning processes and describes how this framework can be used to assess the impact on learning of traveller information systems. The third section describes a laboratory experiment that was carried out to obtain the data necessary to estimate simple learning models based on this framework. The experiment is based on a single link network and focuses on the updating of travellers' departure time decisions. The fourth section discusses the estimation issues that are raised by the model. These are made unusually complex by the need to take account both of serial correlation (due to repeated observation being made on the same individuals) and various sources of heteroscedasticity which arise due to the learning process itself. The fifth section presents the estimation results and discusses their significance. The final section presents some overall conclusions from the work and identifies key questions for future research.

EXISTING APPROACHES TO MODELLING TRAVELLER LEARNING

In this chapter we interpret the notion of learning as referring to the process whereby travellers' expectations of future travel conditions are formed and modified as a result of their personal experience of using the transport system. The dynamics of these expectations are clearly related to but are also quite distinct from the dynamics of expressed travel behaviour. The dynamics of travel behaviour will of course be influenced by travellers' changing expectations regarding system performance, but also by a wide variety of other factors unrelated to expectation formation (e.g., changing needs and obligations or changing external constraints on behaviour). We believe it is important to distinguish separately the influence of these different factors and it is for this reason that we adopted our particular interpretation of learning.

One of the first authors to explicitly consider the issue of traveller learning was Horowitz (1984), who proposed a number of alternative models to describe how travellers update their expectations of network costs in the context of a simple route choice problem involving two parallel routes. The simplest model proposed by Horowitz envisages that travellers maintain a memory of all previous experienced travel times and formulate their expectations for the current time period as a weighted average of these previous experiences. Thus if $\hat{c}_{i,t}$ is the expected travel time for a particular traveller i time period t and $c_{i,k}$ is the actual travel time experienced in time period k, Horowitz's updating rule is:

$$\hat{c}_{i,t} = \sum_{k=1}^{t-1} w_k c_{i,k} + \epsilon_{i,t} \tag{1}$$

where $\epsilon_{i,t}$ is an identically and independently distributed error, and the weights w_k satisfy $\sum_{k=1}^{t-1} w_k = 1$. This model (which is formally identical to a distributed lag model) states that current expectations depend both upon previous experience and a 'learning error', where the latter is conceived as a simple random perturbation affecting the computation of the updating process. Horowitz suggested a variant of this model in which the expectation of travel time in the current time period is formed on the basis of previous *subjective* perceptions rather than previous objective experience:

$$\hat{c}_{i,t} = \sum_{k=1}^{t-1} w_k(c_{i,k} + \epsilon_{i,k}) \tag{2}$$

where the terms $(c_{i,k} + \epsilon_{i,k})$ represent the traveller's subjective perception of the experienced travel time $c_{i,k}$ in time period k and $\epsilon_{i,k}$ is an iid error. Note that in this model, error is conceived as affecting the process of travel time *perception*, rather than the updating process. Clearly, it would possible to combine the representation of both these sources of error in a single model (although this was not stated explicitly by Horowitz).

Both the learning models suggested by Horowitz imply that travellers possess a potentially unlimited memory (and corresponding weight set), which is cognitively implausible (Anderson, 1995). An obvious generalisation is therefore to restrict the size (length) of the memory to some finite duration. Polak and Hazelton (1998) proposed a learning model of this type in which the length of memory is restricted to τ time periods and the weights w are parameterised according to a geometrically declining model. Their model took the form:

$$\begin{aligned} \hat{c}_{i,t} &= \sum_{k=t-\tau}^{t-1} w_k c_{i,k} + \epsilon_{i,t} \\ w_k &= \lambda^k \Lambda \end{aligned} \tag{3}$$

where λ ($0<\lambda\leq 1$) is a measure of the rate at which past experience is discounted in the formation of current expectations (the smaller the value of λ the more rapid is the discounting of past experience), Λ is a scaling coefficient to ensure that the weights sum to unity and ϵ_t is an iid error. As far as we are aware, no attempts have yet been made to estimate empirical parameters of any of the above models, although this would be entirely feasible for the Polak-Hazelton model (at least for moderate values of τ).

The most frequently encountered learning model in the literature is the exponentially weighted moving average model (e.g., Axhausen *et al.*, 1995; Ben-Akiva *et al.*, 1991, Emmerink, 1996; 1998; van Berkum and van der Mede, 1993; Vaughn *et al.*, 1993). In the present context, this model takes the form:

$$\hat{c}_{i,t} = \hat{c}_{i,t-1} + \theta(c_{i,t-1} - \hat{c}_{i,t-1}) \tag{4}$$

with the single smoothing parameter θ ($0 \leq \theta \leq 1$). Under this model, travellers are regarded as updating their expectation of travel time on the basis of the difference between the experienced and the expected travel time in the pervious time period. It is formally equivalent to maintaining a memory of the exponentially weighted moving average of past travel time experiences. The parameter θ effectively measures the speed with which expectations adapt to experience. We are aware of only two studies that have attempted to estimate empirically the value of this speed of learning parameter.

Iida, *et al.* (1992) proposed and empirically estimated two related models of traveller learning, using data from a laboratory experiment in which respondents were asked to state their travel time expectations for each of two routes. The first model the authors considered takes the form:

$$\hat{c}_{i,t} = \alpha + c_{i,t-1} + \theta(c_{i,t-1} - \hat{c}_{i,t-1}) + \varepsilon_{i,t} \quad (5)$$

where $\hat{c}_{i,t}$ is the expected travel time in period t for traveller i, $c_{i,t-1}$ is the actual travel time experienced by traveller i in time period t-1, $\varepsilon_{i,t}$ is a random error term which is assumed to be autocorrelated within each individual over time and contemporaneously correlated across all individuals in the same network system, and α and θ are parameters to be estimated. This model is of particular interest for a number of reasons. First, it makes explicit (through the parameter α and the error term $\varepsilon_{i,t}$) the idea that the learning process itself can be subject to biases. This is an important notion which we will discuss in more detail in the next section. And second, it was the first model to explicitly accommodate an autocorrelated error structure. In their empirical work Iida *et al.*, estimate the model by means of a two-stage least squares regression and report values of θ of around 0.5 but also report considerable variation in θ according to respondent's willingness to change route.

The second model proposed and estimated by Iida, *et al.* takes the form:

$$\hat{c}_{i,t} = \alpha + c_{i,t-1} + \theta_1(c_{i,t-1} - \hat{c}_{i,t-1}) + \theta_2(c_{i,t-2} - \hat{c}_{i,t-2}) + \theta_3(c_{i,t-3} - \hat{c}_{i,t-3}) + \varepsilon_{i,t} \quad (6)$$

with the variables and parameters as defined above. With this extended specification, Iida *et al.* found that the leverage of previous experience generally declined rapidly over time (i.e., in their second model, $|\theta_1| >> |\theta_2|$ and $|\theta_1| >> |\theta_3|$). In both models, the estimated values of α were generally significantly different from zero, indicating the presence of certain learning biases.

The study by Axhausen *et al.*(1995) also used data generated by laboratory experiments to estimate a learning model based on the form given in equation (4). Unlike Iida *et al.,* however, the models presented by Axhausen *et al.* were estimated separately on data for each individual, rather than on data pooled across individuals. This permitted the authors to explore the issue to heterogeneity in learning. The results indicate that there exists considerable variation in learning speed across individuals, with a significant proportion of the sample clustered at or near the point θ = 0 and a smaller but still sizable proportion clustered close to θ = 1 and with relatively little mass between these two points. This suggests a rather stark segmentation in the population between those who largely ignore recent experience and those who are largely guided by it. Some caution is, however, needed in the interpretation of these results since the model estimation appears to have been performed using a non-standard sum of squares criterion and it is unclear for example what specific assumptions were made regarding error structures. Moreover, the number of observations available for the estimation of models at the level of the individual was small, limiting the precision with which the parameters θ could be estimated.

An alternative approach was proposed by Mahmassani and Chang (1986) who specified a model in which travellers were assumed to update their expectations regarding future travel conditions on the basis of their arrival time at their destination. In the current context, their model took the form:

$$\hat{c}_{i,t} = c_{i,t-1} + \alpha_i \delta^E_{i,t-1} SD_{i,t-1} + \beta_i \delta^L_{i,t-1} SD_{i,t-1} + \varepsilon_{i,t} \tag{7}$$

where $SD_{i,t-1}$ is the schedule delay (Small, 1982) experienced by traveller i on day (t-1), $\delta^E_{i,t-1} = 1$ for early arrivals (and zero otherwise) and $\delta^L_{i,t-1} = 1$ for late arrivals (and zero otherwise). The parameters α_i and β_i measure the contribution of early and late schedule delay to the updating of travel time perception. In subsequent work (e.g., Chang and Mahmassani, 1988), Mahmassani and colleagues extended this basic specification in a number of directions, including accommodating autocorrelation and heteroscedasticity in the error term (both of which were found to be empirically significant).

More recently, Jha *et al.* (1998) have proposed a multi-stage Bayesian approach to modelling the updating of perceptions through personal experience and provided information. They specify separate Bayesian updating models for the effects on expectations of pre-trip travel information and post-trip experience and a rule to combine experiential and non-experiential sources of expectation.

Fuji and Kitamura (2000) propose a model (similar in form to the moving average model) in which anticipated travel times are updated on the basis of both experienced travel times and information acquired from various sources including the mass media, telephone and word-of-mouth. Their empirical results suggest that use of non-experiential sources of information reduces reliance on past experience in the formulation of future expectations.

This brief review has highlighted a number of important features in the existing literature on traveller learning. There have been essentially two types of model proposed; the distributed lag formulation and the moving average formulation. Of these two, the moving average model is much more parsimonious, embodies a more plausible updating mechanism and has been the focus of empirical work to date. This empirical work has been limited and in many respects inconclusive. However, this work has pointed to the need to develop much better treatments of inter-personal heterogeneity and to accommodate a variety of errors and biases that appear to affect perception and learning processes.

MODELLING FRAMEWORK FOR TRAVELLER LEARNING

Psychological Insights

The psychological literature abounds with competing and sometimes contradictory definitions and interpretations of the concept of learning (Atkinson *et al.*, 1983; Anderson, 1995). However, it is possible to extract a number of important insights that can usefully influence the development of a model of travel time learning (see Oladeinde (2000) for a more extensive discussion).

It is interesting to note that the moving average model can be given a plausible psychological interpretation in terms of what is called the difference reduction heuristic (Anderson, 1995; Newell and Simon, 1972). This asserts that learning can be thought of as the attempt by individuals to iteratively reduce the discrepancy between their expected and their actual experience: however, the psychological literature also points to a number of weaknesses in the simple moving average model, as stated in equation (4). First, there is a substantial body of evidence (see e.g., Kahneman *et al.*, 1982) which underlines that human judgement and decision making is error-prone, due to, for example, information deficits and cognitive limitations. This suggests that equation (4) needs to be modified by means of the inclusion of an explicit error term to account for learning errors. Moreover, the psychological literature also suggests that such learning errors may vary in magnitude very substantially, according to the learning context and experience. In the context of travel time learning, this suggests that the rather simple (generally homoscedastic) error structures assumed in existing models may need to be replaced with ones that support much more general patterns of heteroscedasticity and autocorrelation.

A further insight from the psychological literature is that individuals often resort to heuristics in order to simplify complex decision making contexts (Kahneman and Tversky, 1973) which in turn can lead to systematic biases in learning outcomes. This suggests that in addition to accommodating a richer set of random error components, an improved model of traveller learning should also accommodate the possibility of systematic biases.

Modelling Framework

We will consider a very simple context in which travellers are assumed to face a commuting journey with a pre-specified required arrival time. The travel time from home to work is assumed to be characterised by a stationary stochastic process, with probability density function $f(\mu_x,\sigma_x)$, where the mean (μ_x) and standard deviation (σ_x) of the distribution are initially unknown to the traveller.

Travellers have the task of choosing a departure time that ensures an acceptable arrival time at work. Assume that on day i a particular traveller departs at time t_i and experiences an actual travel time of x_t (which is a realisation from the distribution defined by f). Define $v_t = PAT - DT_t$, where PAT is the traveller's preferred arrival time. The quantity v_i is the time the traveller allocates to the journey on day t, taking into account the preferred arrival time and current state of knowledge of f. We regard v_t as embodying the traveller's expectation of the travel time. If $v_t < x_t$ the traveller will arrive late and if $v_t > x_t$ the traveller will be early. Interest is focussed on how the quantity v_t evolves over time and on the relationship between v_t, x_t and the characteristics of the pdf $f(\mu_x,\sigma_x)$. Travellers are assumed to form a subjective distribution of travel time which is updated with experience. Let v_t be the mean of the subjective distribution on day t. The learning process involved in the allocation of travel time for the journey to work is treated as the process of learning the characteristics of the travel time distribution from experience.

The pattern of decision making in a one-link network is shown in Figure 1. In the model, the origin, destination and preferred arrival time are assumed known and fixed. The model assumes that travellers start with an initial perception of travel time on the first day. As they make repeated daily trips, their perception of travel time becomes modified. Based on their historical perception of travel time, drivers formulate expectations of travel time for the future. As a result of their expectation regarding future travel time, travellers allocate an amount of travel to the trip. The departure time choice follows as a result of the time allocated for the trip.

FIGURE 1 Framework for Travel Time Learning in a Single Link Network.

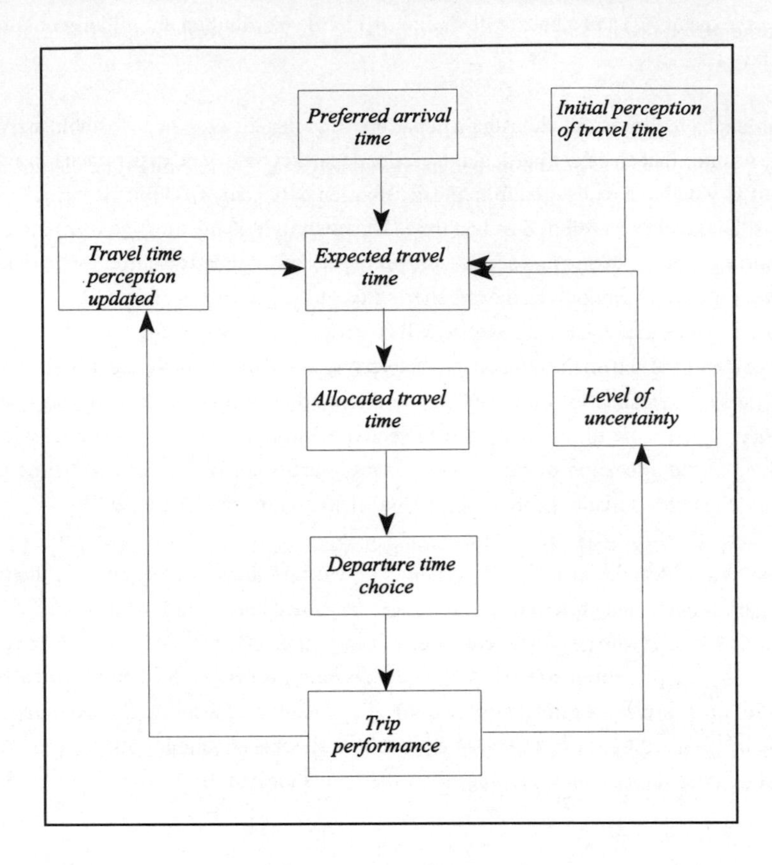

The Basic Learning Model

The basic model we propose for updating the allocated travel time for a regular trip is a modified version of the moving average model:

$$v_{i,t+1} = \alpha + v_{i,t} + \beta_{t+1}(x_{i,t} - v_{i,t}) + \varepsilon_{i,t+1} \tag{8}$$

where

$$\beta_{t+1} = \beta + \omega_{t+1} \tag{9}$$

and $v_{i,t+1}$ and $v_{i,t}$ are the time allocated by individual i to the journey to work in time period t+1 and time period t respectively, α is a learning bias, β_{t+1} is the learning speed in time period t+1 which takes a value in the closed interval [0,1], $x_{t,i}$ is the travel time actually experienced by individual i in time period t (which is assumed to be log normally distributed) and $\varepsilon_{i,t+1}$ is the learning error, which is assumed to be correlated with $x_{t,i}$. The potential for correlation between $\varepsilon_{i,t+1}$ and $x_{i,t}$ is introduced to accommodate the possibility that learning might be disproportionately affected by extreme travel time experiences. The speed of learning parameter β is assumed to be potentially varying in magnitude over time to account for the possibility that learning may become more (or less) efficient over time. The errors ω_{t+1} and ε_{t+1} are assumed uncorrelated and to follow a normal distribution with mean zero and variances σ_{ω}^{2} and σ_{ε}^{2} respectively.

To account for the effects of repeated decision making process involved in the day to day learning, the learning error term is assumed to follow a first order autoregressive process:

$$\varepsilon_{i,t+1} = \rho\varepsilon_{i,t} + \theta_{i,t+1} \tag{10}$$

where ρ is the autocorrelation coefficient to be estimated, $E(\theta_{i,t+1}) = 0$ and $E(\theta_{i,t+1}^{2}) = \sigma_{\theta_i}^{2}$.

We can re-express equation (8) as follows:

$$v_{i,t+1} = \alpha + (1-\beta_{t+1})v_{i,t} + \beta_{t+1}(\mu_x + \Delta_t) + \varepsilon_{i,t+1} \tag{11}$$

where Δ_t is the sampling error from the travel time distribution (i.e., $\text{Var}(\Delta_t) = \sigma_x^2$). This expression can be rearranged to give:

$$v_{i,t+1} = \alpha + (1-\beta_{t+1})v_{i,t} + \beta_{t+1}\mu_x + [\beta_{t+1}\Delta_t + \varepsilon_{i,t+1}] \tag{12}$$

where the term $v_{i,t+1} = (1-\beta_{t+1})v_{i,t} + \beta_{t+1}\mu_x$ represents what we shall term the 'perfect learning model' and the term $[\beta_{t+1}\Delta_t + \varepsilon_{i,t+1}]$ is a compound error term which we assume that travellers are trying to minimise. The term 'perfect learning' is used because under the model $v_{i,t+1} = (1-\beta_{t+1})v_{i,t} + \beta_{t+1}\mu_x$, the traveller's subjective assessment of mean travel time, $v_{t+1,i}$, will always tend asymptotically to the true mean μ_x at a rate that is determined by the parameters β_{t+1}.

In the absence of any systematic learning bias (i.e., if α=0) the degree to which the traveller achieves 'perfect learning' is determined by the interaction between the day-to-day sampling error and the learning error. If learning errors act so as to offset the effect of day-to-day sampling variation then learning will be improved, whereas if learning errors reinforce day-to-day variation the quality of learning will be degraded.

To obtain the fully specified form of the model, we substitute equations (9) and (8) into equation (8):

$$v_{i,t+1} = \alpha + \rho v_{i,t} + (v_{i,t} - \rho v_{i,t-1}) + \beta[(x_{i,t} - v_{i,t}) - \rho(x_{i,t-1} - v_{i,t-1})] + \Omega_{i,t+1} \quad (13)$$

where the fully specified compound error term is given by:

$$\Omega_{i,t+1} = \theta_{i,t+1} + \omega_{i,t+1}(x_{i,t} - v_{i,t}) - \rho\omega_{i,t}(x_{i,t-1} - v_{i,t-1}) \quad (14)$$

which has mean zero and variance:

$$\sigma_{\Omega}^2 = \sigma_{\theta}^2 + \sigma_{\omega}^2[(x_{i,t} - v_{i,t})^2 + \rho^2(x_{i,t-1} - v_{i,t-1})^2] \quad (15)$$

and Corr $(x_{i,t}, \Omega_{i,t+1}) = m$.

As discussed above, m can be viewed as a measure of the quality of travellers' learning. In the learning process, $m > 0$ implies that drivers over-react to large deviations in experienced travel time from the usual experienced travel time. $m = 0$ implies that drivers are insensitive to large deviation in experienced travel time while $m < 0$ implies that drivers are aware of the large deviation in travel time but do not over react to these deviation probably due to vast experience of travelling on the network system or an understanding of the dynamics of the network system.

If inter-personal heteroscedasticity is suspected, Glesjer's test (Greene, 1991) can be used to test if it is present. The specific form adopted for the test in this context is:

$$\sigma_{\theta}^2 = \sigma_{\theta*}^2 \exp(\kappa(x_{i,t} - v_{i,t})^2) \quad (16)$$

Where κ is a parameter to be estimated and $\sigma_{\theta*}^2$ is the variance of the learning model after heteroscedasticity as a result of cross-sectional variation has been removed. If heteroscedasticity due to cross-sectional data is present, then κ will be significantly different from zero, otherwise the error structure is homoscedastic.

Suppose it is assumed that,

$$\sigma_{\omega}^2 = \lambda \sigma_{\theta*}^2 \quad (17)$$

Substituting equation (16) and (17) into equation (15) the error learning variance term becomes:

$$\sigma_{\Omega}^2 = \sigma_{\theta*}^2[\exp(\kappa(x_{i,t} - v_{i,t})^2) + \lambda B] \quad (18)$$

where,

$$B = [(x_{i,t} - v_{i,t})^2 + \rho^2(x_{i,t-1} - v_{i,t-1})^2] \qquad (19)$$

and λ is a parameter that measures the degree of heterogeneity in learning speed over time.

Modelling the Effect of Provided Information

The effect of descriptive travel time information can be modelled in this framework, by assuming that the provided information is equivalent to a pseudo-experience, which further updates travellers' expectations. The form of the model used is:

$$v^r_{i,t+1} = (1-c_{t+1})v_{i,t+1} + c_{i,t+1}(\pi_{i,t+1}) + \eta_{i,t+1} \qquad (20)$$

where,

$$c_{t+1} = c + \tau_{t+1} \qquad (21)$$

and τ is assumed to be uncorrelated with η.

The term π_{t+1} is the information in descriptive form, supplied to the traveller at the start of time period t+1 (but before the journey commences), c_{t+1} is the 'credibility' of the information from the traveller's point of view in time period t+1, c is the average credibility travellers attach to traffic information over the whole study period. The value of the credibility parameter reflects the weight given to provided traffic information relative to experience. If the traveller perceives the traffic information as worthless, then the value zero is assigned to the credibility of the traffic information. If the traveller believes the information system unquestioningly, then the value of the credibility parameter is unity.

Substituting equation (21) in (20) gives:

$$v^r_{i,t+1} = (1-c)v_{i,t+1} + c\pi_{i,t+1} + \zeta_{i,t+1} \qquad (22)$$

where,

$$\zeta_{i,t+1} = \tau(\pi_{i,t+1} - v_{i,t+1}) + \eta_{i,t+1} \qquad (23)$$

is the compound information updating error term which is assumed to follow a normal distribution with mean zero and variance:

$$\sigma^2_\zeta = \sigma^2_\eta + \sigma^2_\tau(\pi_{i,t+1} - v_{i,t+1})^2 \qquad (24)$$

Hence the compound information updating error variance is heteroscedastic over time as a result of learning from the traffic information.

If cross-sectional heteroscedasticity is suspected, it can be parameterised in a manner similar to that used in the basic learning model. This implies that we replace σ^2_η with

$$\sigma^2_\eta = \sigma^2_{\eta*}.\exp(\chi(\pi_{i,t+1} - v_{i,t+1})^2) \tag{25}$$

where χ is a parameter that measures cross-sectional heteroscedasticity and $\sigma^2_{\eta*}$ is the variance of the information learning model after heteroscedasticity as a result of cross-sectional variation has been removed.

This leads to the following expression

$$\sigma^2_\zeta = \sigma^2_{\eta*}.[\exp(\chi(\pi_{t+1} - v_{t+1})^2) + \psi(\pi_{t+1} - v_{t+1})^2] \tag{26}$$

where ψ is a parameter that measures the heteroscedasticy over time in the information updating process.

LABORATORY EXPERIMENT

In order to collect information to enable this model system to be estimated, a series of laboratory experiments were carried out. Full details of the design and execution of these experiments are given in Oladeinde (2000). Only a brief summary will be given here.

Sixty three participants took part in two simulation experiments carried out in October 1998 at Imperial College. The first experiment involved participants making travel time decisions in a one route network and the second experiment involved the same participants making travel time decisions in a two route network connected by a single origin and destination point. This chapter concentrates on the results from the first experiment.

The participants were divided into five groups with a different level of information accuracy provided in each group in order to facilitate the decision making tasks (the fifth group receives no information). The four separate groups have 14, 14, 15 and 20 participants respectively with each different level of information accuracy.

Group 1: This group receives no information.

Group 2: This group received good quality information throughout the simulation experiment.

Group 3: This group received information that was initially bad but which got better as the simulation progressed.

Group 4: This group received information that was initially good but which got worse as the simulation progressed.

Group 5: This group received information that was initially bad but which then got better and then worse as the simulation progressed.

Participants were informed that they arrived at the place of work no later than 9:00 A.M. Initially, the average travel time in both experiments were given to each participant. In both experiments, travel time was modelled as a lognormal distribution. In the first experiment, which involves only one route, participants were informed that the average travel time on the route is 37 minutes. In the second experiment, the two routes were labelled route 1 and route 2. Participants were informed that route one is a dual carriage freeway which is 20km long. Average travel time on this route is 42 minutes and travel time can vary quite significantly from day-to-day. Route two is also a dual carriageway, which runs through a busy town centre. It is 14.5km long and average travel time on this route is 38 minutes. Travel time can also vary quite significantly from day-to-day on this route.

In both experiments, participants first feed in their departure time choice in the first experiment and their route choice and departure time choice in the second experiment. Based on these choices, predictive travel time information were given to each participant on the available routes with the opportunity of adjusting their choices. Both experiments subjected participants to 15 simulation days.

MODEL ESTIMATION ISSUES

As a result of the complex structure of the error term in the allocation and information learning models, the computation of maximum likelihood parameter estimates was somewhat involved.

Likelihood Function

The key requirement for the estimation of both models is to evaluate the joint distribution of the compound error terms. In the case of the basic learning model, we model the distribution of travel time as a lognormal distribution and we assume that the conditional distribution of the learning error term is assumed to follow a normal distribution. The joint distribution of the most recently experienced travel time and the learning error term is obtained as follows:

$$f(\Omega_{i,t+1}, x_{i,t+1}) = f(\Omega_{i,t+1}|x_{i,t}) f(x_{it}) \tag{27}$$

where,

$$f(\Omega_{i,t+1}|x_{i,t}) = \frac{1}{\sqrt{2\pi}\sigma_\Omega\sqrt{(1-m^2)}} \exp -\frac{1}{2}\left[\left(\frac{\Omega_{i,t+1} - m\left(\frac{\sigma_\Omega}{\sigma_x}\right)(x_{i,t}-\mu_x)}{\sigma_\Omega\sqrt{(1-m^2)}}\right)^2\right] \tag{28}$$

and

$$f(x_{i,t}) = \frac{1}{x_{i,t}\sqrt{2\pi}\sigma} \exp -\frac{1}{2}\left[\left(\frac{\ln x_{i,t} - \mu}{\sigma}\right)^2\right] \tag{29}$$

with μ and σ^2 being the mean and variance of the lognormal travel time distribution, m is the correlation coefficient between the learning error term and the most recently experienced travel time and the rest of the terms are as defined in the previous section.

The likelihood function denoted by $LF(\beta,m,\rho,\kappa,\sigma_{\theta^*})$ can be written as follows (where the summation extends over all time periods and individuals)

$$LF(\beta,m,\rho,\kappa,\sigma_{\theta^*}) = \prod_{i,t} f(\Omega_{i,t+1}|x_{i,t}) f(x_{i,t}) \tag{30}$$

The log-likelihood function $\mathfrak{L}_a$ is thus:

$$\mathfrak{L}_a = Log(LF(\beta,m,\rho,\kappa,\sigma_{\theta^*})) = \sum_{i,t} Log(f(\Omega_{i,t+1}|x_{i,t}) f(x_{i,t})) \tag{31}$$

For the estimation of the information updating model, the error term in equation (21) was assumed to follow a normal distribution, thus:

$$f(\zeta_{t+1}) = \frac{1}{\sqrt{2\pi}\sigma_\zeta} \exp -\frac{1}{2}\left(\frac{\zeta_{t+1}}{\sigma_\zeta}\right)^2 \tag{32}$$

The parameters of the information learning model to be estimated are c,χ,ψ and $\sigma^2_{\eta^*}$

The likelihood function denoted by $LF(c,\chi,\psi,\sigma_{\eta^*})$ is written as follows,

$$LF(c,\chi,\psi,\sigma_{\eta^*}) = \prod_{i,t} f(\zeta_{i,t+1}) \tag{33}$$

The log-likelihood function $\mathcal{L}_c$ is thus:

$$\mathcal{L}_c = \log(LF(c,\chi,\psi,\sigma_{\eta'})) = \sum_{i,t} \log(f(\zeta_{i,t+1})) \tag{34}$$

Goodness of Fit

The overall goodness-of-fit of the models for both experiments was carried out using the Theil's inequality coefficient (Theil, 1961). This coefficient is defined as:

$$U = \frac{\sqrt{\frac{1}{T}\sum_{i=1}^{T}(v_{i,t}^{s}-v_{i,t}^{a})^2}}{\sqrt{\frac{1}{T}\sum_{i=1}^{T}(v_{i,t}^{s})^2}+\sqrt{\frac{1}{T}\sum_{i=1}^{T}(v_{i,t}^{a})^2}} \tag{35}$$

where U is the Theil inequality coefficient which takes values in the closed interval [0,1], $v_{i,t}^{a}$ is the actual allocated travel time given by a participant i in time period t, $v_{i,t}^{s}$ is the predicted value of the allocated travel time and T is the number of observations. If $U = 0$, $v_{i,t}^{s} = v_{i,t}^{a}$ for all t and there is a perfect fit. Values of U less than 0.2 are generally regarded as representing a good fit. A more extensive discussion of goodness of fit measures is given in Oladeinde (2000).

Estimation Results

Table 1 summarises thc key parameters in the model system and Tables 2 and 3 show the estimated values of the parameters of the learning models for all the groups in experiment 1 before and after the provision of traffic information. The numbers in parentheses are t-statistics.

Considering first the results shown in Table 2. The value of U for all groups is less than 0.1, indicating that the models fit the data extremely well. The value of the average learning speed parameter (β) is in the range 0.1 to 0.25 for all groups, with the exception of group 2, where it is somewhat lower. It is strongly significant in all cases. These magnitudes are broadly similar to those found by Iida *et al.*, (1992) and rather different from those reported in Axhausen *et al.* (1995).

TABLE 1 Summary of Key Model Parameters.

Parameter	Definition
λ	Temporal heteroscedasticity parameter
β	Learning parameter
m	Quality of learning parameter
ρ	Autocorrelation parameter
κ	Cross-sectional Heteroscedasticity in basic learning model
$\sigma_{\theta*}$	Standard deviation of learning error term
$\sigma_{\eta*}$	Standard deviation of information learning error term
c	Information credibility parameter
χ	Cross-sectional heteroscedasticity in the presence of traffic information
ψ	Temporal heteroscedasticity parameter as a result of traffic information

Significant temporal heteroscedasticity (λ) in the learning speed is present in 3 of the five groups, with Groups 3 and 4 being distinctly different from the rest. This result constitutes reasonably strong evidence that in the absence of externally provided information, the learning speed of a given individual will be subject to temporal variations. Significant residual autocorrelation (ρ) is also present in 3 of the five groups. This suggest that although the model captures certain aspects of the dynamics of the learning process, there remain other unobservable factors affecting the speed of traveller learning. Inter-personal heteroscedasticity (κ) is strongly present in 3 of the five groups, suggesting that there are indeed significant person-specific factors affecting the learning process, net of the influence of learning dynamics. The results also indicate that learning errors ($\sigma_{\theta*}$) are strongly present and of broadly similar magnitude in all groups.

The quality of learning parameter (*m*) was of statistical insignificance in Groups 1, 2 and 5, indicating that for these groups, there was not significant correlation between learning and day-to-day sampling errors. In group 4 the estimated value of *m* is statistically significant and positive, indicating that travellers in this group tended to under-react to large deviations in travel time, thus improving the overall quality of their learning. By contrast, individuals in group 3 tended to over-react to large deviations in travel time. This may in part explain some of the otherwise anomalous results obtained for this group.

TABLE 2 Parameter Estimate from Experiment 1: Prior to Traffic Information being Supplied.

	Group 1	Group 2	Group 3	Group 4	Group 5
λ	0.0051 (6.697)	0.0108 (5.326)	See note	-0.00003** (-0.057)	0.0080 (4.927)
β	0.388 (6.801)	0.197 (3.587)	0.311 (5.221)	0.158 (3.889)	0.276 (5.253)
m	-0.121** (-1.340)	-0.003** (-0.028)	-0.197 (-1.998)	0.146 (2.082)	-0.042** (-0.583)
ρ	-0.049 (-0.869)	-0.340 (-7.691)	-0.097** (-1.009)	-0.366 (-7.190)	-0.299 (-5.786)
κ	-0.458 (-6.207)	-0.317 (-5.883)	-0.002 (2.475)	See note	-0.022 (-3.850)
$\sigma_{\theta*}$	6.600 (20.711)	4.995 (14.620)	4.985 (33.181)	4.539 (22.862)	4.569 (23.494)
U	0.05	0.06	0.08	0.07	0.07

** insignificant at the 95% confidence level

* significant at the 90% confidence level

Note: In the initial estimation, the estimated value of λ for Group 3 was negative and statistically significant. The model for this group was re-estimated with λ constrained to zero. Similarly, the initially estimated value of κ in Group 4 was positive and significant and was treated likewise. See Maddala (1977) for a comprehensive discussion of the interpretation and treatment of this type of problem.

Considering the results in Table 3, the overall goodness of fit for all groups (as indicated by Theil's U) is excellent and there are also some interesting changes apparent once external information is provided. In all groups the learning heteroscedasticity parameters (λ and κ) and the learning autocorrelation parameter (ρ) become essentially insignificant. The presence of 'within day' traffic information seem to have reduced the sources of difference in learning both within individuals over time and between different individuals. The average speed of leaning (λ) tends to increase (to between 0.4 to 0.7) and the quality of learning (m) improves. There is little change to the magnitude of the learning error ($\sigma_{\theta*}$) and no significant difference in its magnitude across the different information conditions.

Turning to the parameters characterising the information updating model, we observe some interesting patterns across the different information conditions. The information credibility parameter (c) is positive and statistically significant in all information conditions, indicating that the provided information significantly influences the learning process. As one might expect, the highest value of c arises in circumstances where high quality information is provided throughout (Group 2) and significantly lower values of c are evident in Groups 3 and 4 where information is of variable quality (credibility is roughly halved). Surprisingly, the value of the credibility parameter associated with group 5 (where information is initially bad, then improves and then reverts) is close to the value observed in Group 1. There is significant temporal heteroscedasticity (ψ) in the information updating across all groups. The error in information updating ($\sigma_{\eta*}$) is significant in all information conditions, but significantly smaller when the provided information is of good quality (Group 1).

TABLE 3 Parameter Estimate from Experiment 1: After Traffic Information was Supplied.

	Group 2	Group 3	Group 4	Group 5
λ	0.0032^{*} (1.619)	0.0001^{**} (0.175)	0.0036^{**} (1.532)	-0.0003^{**} (-0.387)
β	0.451 (5.920)	0.451 (5.920)	0.566 (8.507)	0.691 (13.933)
m	-0.542^{**} (-11.926)	-0.542 (-11.926)	-0.453 (-8.994)	-0.547 (-13.184)
ρ	-0.072^{**} (-1.388)	-0.078^{**} (-1.328)	-0.0416 (-0.708)	-0.030^{**} (-0.610)
κ	Not present	Not present	Not present	Not present
$\sigma_{\theta*}$	5.153 (22.090)	5.299 (22.741)	5.441 (18.709)	5.895 (22.148)
$\sigma_{\eta*}$	1,576 (13.033)	4.312 (18.164)	2.255 (17.327)	4.187 (17.715)
χ	Not present	Not present	Not present	Not present
c	0.504 (11.711)	0.244 (6.434)	0.316 (9.553)	0.466 (14.894)
ψ	0.097 (3.895)	0.0027 (3.476)	0.0193 (4.118)	0.0044 (2.557)
U	0.032	0.057	0.073	0.067

** insignificant at the 95% confidence level
* significant at the 90% confidence level

CONCLUSIONS

In this chapter we have developed a simple theoretical framework and model of traveller learning in the context of uncertain travel times. This framework and model builds on the existing literature, but extends it in a number of important respects, both conceptually and methodologically. The models have been estimated using data from a laboratory experiment and have produced results that are broadly plausible but which also quantify and give insight into the relationship between experience, provided information and traveller learning.

The are a number of directions in which the current work could usefully be extended. Although the context in which we have developed the current model is that of a destination-constrained departure time choice, we have not given detailed attention to the effects of trip scheduling considerations (e.g., Small, 1982) on learning. It would be possible to extend the current framework to deal with these effects and indeed to formally link the learning/updating model to this and other dimensions of travel choice. Another potentially fruitful direction would be to relax the assumption that the travel time distribution is stationary in time. In the first instance this could be done by assuming that the mean travel time experience as a result of a departure at time *DT* $\mu_x(DT)$ is a simple function of *DT* and that at each level of travel time stochastic variation takes place on a day-to-day basis about $\mu_x(DT)$. The objective of learning would then be to learn the shape of $\mu_x(DT)$.

REFERENCES

Anderson, J.A.,(1995) *Learning and Memory: An Integrated Approach*. John Wiley and Sons, New York.

Atkinson R.L., Atkinson R.C. and Hilgard E.R. (1983) *Introduction to Psychology*, Eighth edition, Harcourt Brace Jovanovich Publishers.

Axhausen, K.W., Dimitrakopoulou, E. and Dimtropoulos, I. (1995) 'Adapting to change: Some evidence from a simple learning model', Proceedings PTRC Summer Annual Meeting.

Ben-Akiva, M.E., de Palma, A., and Kaysi, I. (1991) 'Dynamic network models and driver information systems', *Transportation Research* **25A**(5) 251-266.

Chang, G-L. and Mahmassani, H.S. (1988) 'Travel time prediction and departure time adjustment behaviour dynamics in a congested traffic system', *Transportation Resreach* **22B**(3) 217-232.

Emmerink, R.H.M., (1996) *Information and Pricing in Road Transport*. Ph.D Dissertation, Tinbergen Institute Research series, Vrije Universiteit Amsterdam

Fuji, S. and Kitamura, R. (2000) 'Anticipated travel time, information acquisition and actual experience: The case of Hanshin Expressway Route Closure' paper presented to the 79th Annual Meeting of the Transportation Research Board.

Green, W.H. (1995) *Economic Analysis*, John Wiley and Sons, Chichester.

Hazelton, M., Lee, S. and Polak, J.W. (1996) 'Stationary states in stochastic process models of traffic assignment: A Markov Chain Monte Carlo approach' **in** J-B. Lescot (Ed.) *Transportation and Traffic Theory*, Pergamon, Oxford.

Horowitz A.J. (1984) 'The stability of stochastic equilibrium in a two-link transportation network', *Transportation Research* , **18B** (1) 13-28.

Iida Y., Akiyama T., and Uchida T. (1992) 'Experimental analysis of dynamic route choice Behaviour', *Transportation Research* , **26B** (1) 17-32.

Jha, M., Madanat, S. and Peeta, S. (1998) 'Perception updating and day-to-day travel choice dynamics in traffic networks with information provision' *Transportation Research* , **6C** (3) 189-212.

Kahneman, D. and Tversky, T. (1973) 'On the psychology of prediction', *Psychological Review* **80** 237-251.

Kahneman, D., Slovic, P. and Tversky T. (eds) (1982) *Judgement under Uncertainty: Heuristics and Biases*, Cambridge University Press, Cambridge.

Koutsopoulos, N.H. and Xu, H. (1993) 'An information discounting routing strategy for advanced traveller information systems' *Transportation Research* , **1C** (3) 249-264.

Mahmassani, H.S. and Chang, G. (1986) 'Experiments with departure time dynamics of urban commuters' *Transportation Research* , **20B** (4) 297-320.

Pindyck, R.S. and Rubinfeld D.L. (1991) *Econometric Models and Economic Forecasts*. McGraw-Hill International Editions.

Oladeinde, F.A. (2000) *A Model of Traveller Learning with Applications to Traffic Information Systems*, PhD Dissertation, Centre for Transport Studies, Department of Civil and Environmental Engineering, Imperial College of Science Technology and Medicine.

Polak, J.W. and Hazelton, M.L. (1998) 'The influence of alternative traveller learning mechanisms on the dynamics of transport systems' Proceedings 26th European Transport Forum, PTRC, London.

Small, K.A. (1982) 'The scheduling of consumer activities: work trips' *American Economic Review* **72**(3) 467-479.

Theil, H. (1971) *Principles of Econometrics*, Wiley, New York.

van der Mede, P. and van Berkum E. (1991) 'BEAST: a Behavioural approach to simulating travellers', paper presented at the 6th International Conference on Travel Behaviour, Quebec.

van Berkum, E and van der Mede, P.,(1993) *The Impact of Traffic Information: Dynamics in Route and Departure Time Choice*. Doctoral Thesis, Proefschrift Technische Universiteit Delft

CHAPTER 3

TRAFFIC ASSIGNMENT WITH STOCHASTIC FLOWS AND THE ESTIMATION OF TRAVEL TIME RELIABILITY

David Watling, Institute for Transport Studies, University of Leeds, U.K.

1. INTRODUCTION

The transport planner is faced with an ever-expanding array of equilibrium traffic assignment models for assessing the performance of a road traffic network. The most well known of these are the deterministic and stochastic user equilibrium (DUE and SUE) models, for both the within-day static (Sheffi, 1985) and within-day dynamic (Ran & Boyce, 1993) cases, although a number of variants exist (e.g. Dial, 1997). A common aspect of all these approaches is that the flow variables are assumed deterministic, and so they are poorly suited to examining the impact on network performance of—say—variability in the inter-zonal trip rates. Although the SUE approach is able to represent perceptual differences across the driver population in the evaluation of a given (actual) travel time, the actual travel time is itself represented as a deterministic quantity. Therefore, such models are unable to provide predictions of travel time variability. It should also be noted that, intuitively, these two lacking features (flow variability and travel time variability) are intimately connected.

In view of such limitations, and with the background of a growing interest in the reliability of a network's performance (Wakabayashi & Iida, 1992; Bell & Iida, 1997), a number of extensions to these modelling approaches have been proposed. A pragmatic, *post-analysis* approach adopted by a number of authors has been to assume that a conventional equilibrium model in some sense represents mean behaviour, and to use Monte Carlo simulation to replicate variations about this mean (Smith & Russam, 1989; Mutale, 1991; Barcelo, 1991; Koutsopoulos & Xu, 1993; Willumsen & Hounsell, 1994; McDonald & Lyons, 1996). In contrast, some authors have proposed *pre-analysis* approaches, in which the link performance functions are modified, to represent independent Poisson link flow variation (Bell, 1991) or stochastic capacity variations (Arnott *et al*, 1991). The modified functions are then used in a conventional model, thus reflecting the potential impact of variability on mean conditions. Finally, a number of new modelling approaches have been proposed. Mirchandani & Soroush (1987) proposed a modified form of SUE in which actual travel costs are stochastic. However, in contrast to the technique proposed later in this chapter, they assume the probability distribution to be exogenously-specified and independent of variations in flows; indeed, flows are still represented as deterministic variables. Perhaps the most radical departure is that championed by Cantarella & Cascetta (1995), where the day-to-day evolution of flows is explicitly modelled as a discrete time stochastic process. Thus flow variations, and their relationship with travel time variations, are automatically represented.

The objective of this chapter is to draw on elements of all these approaches, to devise a modelling procedure which:

— simultaneously estimates mean performance and reliability in a consistent framework;
— relates travel time/cost variability to flow variability, while taking proper account of correlations between links;
— makes maximum use of network data typically available, while minimising the need for new data inputs; and
— possesses a computationally-efficient solution algorithm.

Unlike the approach of Cantarella & Cascetta (1995), however, underlying this development will be a desire to retain as much as possible of the tried-and-tested equilibrium approach to network performance prediction.

The chapter is structured as follows. In section 2, the necessary notation is introduced. This is then used to describe in section 3 a recently-proposed generalised equilibrium approach, which is specifically designed for accommodating stochastic flow and travel time variables. The basic approach, which is able to represent stochastic, day-to-day variation in route choice, and has been extended to represent stochastic demand, is further extended here to represent stochastic capacities, and to provide estimates of travel time variances and covariances. A simplified description of these conditions is presented in section 4. In section 5, a heuristic solution algorithm is described, and the results of applying the method to a simple illustrative example are reported.

2. NOTATION

Define the following notation:

A = number of network links

W = number of inter-zonal movements

N = number of possible routes that pass through a link at most once

$$R_k = \left\{ r + \sum_{j=1}^{k-1} N_j : r = 1,2,\ldots,N_k \right\} = \text{index set of } N_k \text{ possible routes for movement } k$$

$\mathbf{q}$ = W-vector of inter-zonal demand rates per hour

Γ = $N \times W$ matrix of 0/1 entries, with $\Gamma_{rk} = 1$ only if route r relates to movement k

Δ = $A \times N$ matrix of 0/1 entries, with $\Delta_{ar} = 1$ only if link a is part of route r

$\Re_+^n$ = space of n-dimensional non-negative reals

Ω_1 = $\{\mathbf{f} \in \Re_+^N : \Gamma^\perp \mathbf{f} = \mathbf{q}\}$ = convex set of demand-feasible route flow rates

Ω_2 = $\{\mathbf{v} \in \Re_+^A : \mathbf{v} = \Delta\,\mathbf{f}$ where $\mathbf{f} \in \Omega_1\}$ = convex set of demand-feasible link flow rates

$t_a(\mathbf{v})$ = cost of travelling along link a at a given link flow rate vector $\mathbf{v}$

$\mathbf{t}(\mathbf{v})$ = A-vector of functions $t_a(\mathbf{v})$ (a=1,2,…,A)

$\mathbf{c}(\mathbf{f}) = \Delta^\perp \mathbf{t}(\Delta\,\mathbf{f})$ = N-vector of implied route cost *vs.* route flow performance functions

$\mathbf{A}^\perp$ = transpose of the matrix $\mathbf{A}$

$p_r(\mathbf{u})$ = proportion of drivers on movement k who choose route $r \in R_k$ when route costs are $\mathbf{u}$

$\mathbf{p}(\mathbf{u})$ = N-vector of functions $p_r(\mathbf{u})$ $(r = 1,2,\ldots,N)$

τ = length of modelled time period (in hours)

$\tilde{\mathbf{q}}$ = W-vector of *absolute* (integer) inter-zonal demands (N.B. $\mathbf{q} = \tau^{-1}\tilde{\mathbf{q}}$)

Z_+^n = space of n-dimensional non-negative integers

$\tilde{\Omega}_1$ = $\{\tilde{\mathbf{f}} \in Z_+^N : \Gamma^\perp \tilde{\mathbf{f}} = \tilde{\mathbf{q}}\}$ = set of demand-feasible absolute route flows

$\tilde{\Omega}_2$ = $\{\tilde{\mathbf{v}} \in Z_+^A : \tilde{\mathbf{v}} = \Delta\,\tilde{\mathbf{f}}$ where $\tilde{\mathbf{f}} \in \tilde{\Omega}_1\}$ = set of demand-feasible absolute link flows

$\mathbf{V}$, $\mathbf{F}$, $\tilde{\mathbf{V}}$, $\tilde{\mathbf{F}}$, $\mathbf{C}$ and $\mathbf{T}$ = vector random variables of relevant flow/cost vectors $\mathbf{v}$, $\mathbf{f}$, $\tilde{\mathbf{v}}$, $\tilde{\mathbf{f}}$, $\mathbf{c}$ and $\mathbf{t}$

According to Sheffi (1985), the route flow rate vector $\mathbf{f} \in \Omega_1$ is termed a *stochastic user equilibrium* (SUE) if and only if

$$\mathbf{f} = \text{diag}(\Gamma\mathbf{q}).\mathbf{p}(\mathbf{c}(\mathbf{f})) \tag{2.1}$$

Alternatively, the link flow rate vector $\mathbf{v} \in \Omega_2$ is termed a SUE if and only if

$$\mathbf{v} = \Delta.\text{diag}(\Gamma\mathbf{q}).\mathbf{p}(\Delta^\perp \mathbf{t}(\mathbf{v}))\,. \tag{2.2}$$

The typical assumption is that $\{p_r(\mathbf{u}) : r \in R_k\}$ is a random utility model:

$$p_r(\mathbf{u}) = \Pr(u_r + e_r \le u_s + e_s\ ,\ \forall s \in R_k\ ,\ s \ne r) \qquad (r \in R_k;\ k = 1,2,\ldots,W)$$

where $\{e_r : r \in R_k;\ k = 1,2,\ldots,W\}$ follow some given joint probability distribution.

3. EQUILIBRIUM CONDITIONS FOR STOCHASTIC FLOW PROBLEMS

We consider in turn stochastic variation in route choice (section 3.1), demand flows (section 3.2) and capacities (section 3.3), with a model of increasing generality.

3.1 Stochastic variation in route choice

Suppose, then, that there is stochastic, day-to-day variation in the route choices of drivers, but that both demands and capacities are deterministic. Define partitions, $\mathbf{p}(\mathbf{u}) = \begin{pmatrix} \mathbf{p}_{[1]}(\mathbf{u}) \\ \mathbf{p}_{[2]}(\mathbf{u}) \\ \vdots \\ \mathbf{p}_{[W]}(\mathbf{u}) \end{pmatrix}$ and $\tilde{\mathbf{F}} = \begin{pmatrix} \tilde{\mathbf{F}}_{[1]} \\ \tilde{\mathbf{F}}_{[2]} \\ \vdots \\ \tilde{\mathbf{F}}_{[W]} \end{pmatrix}$, of the choice probability function $\mathbf{p}(\mathbf{u})$ and the absolute route flow vector $\tilde{\mathbf{F}}$. In particular, then, assume that *given* a vector of mean perceived link costs $\mathbf{y}$ ('mean' in the sense of averaged across the driver population), for each inter-zonal movement k=1,2,…,W independently, each of the $\tilde{q}_k$ drivers independently chooses between the available routes with probabilities $\mathbf{p}_{[k]}(\Delta^{\perp}\mathbf{y})$. Under such an assumption, the distribution of $\tilde{\mathbf{F}}_{[k]}$, conditional on random link costs $\mathbf{Y}$, is given by

$$\tilde{\mathbf{F}}_{[k]} \,\Big|\, \mathbf{Y} = \mathbf{y} \;\sim\; \text{Multinomial}(\tilde{q}_k, \mathbf{p}_{[k]}(\Delta^{\perp}\mathbf{y})) \qquad \text{(independently for } k = 1,2,\ldots,W\text{)}. \qquad (3.1)$$

The link flows that arise from the stochastic route choice behaviour (3.1) will also be random variables, as will (through the link performance functions) the link travel costs/times. We shall subsequently define equilibrium in this stochastic flow setting as a condition on the vector ψ, of cardinality $|\tilde{\Omega}_2|$ (the number of demand-feasible integer link flow states), whose elements represent the (unknown) probabilities $\{\Pr(\tilde{\mathbf{V}} = \tilde{\mathbf{v}}) : \tilde{\mathbf{v}} \in \tilde{\Omega}_2\}$. That is to say, ψ is a representation of the joint probability distribution of $\tilde{\mathbf{V}}$, the vector of network link flows. This distribution is related to the route flow probability distribution ς (a column vector of probabilities $\{\Pr(\tilde{\mathbf{F}} = \tilde{\mathbf{f}}) : \tilde{\mathbf{f}} \in \tilde{\Omega}_1\}$ of dimension $|\tilde{\Omega}_1|$), by $\psi = \Pi\varsigma$, where Π is a $|\tilde{\Omega}_2| \times |\tilde{\Omega}_1|$ matrix with elements

$$\Pi_{ij} = \begin{cases} 1 & \text{if the route flow } \tilde{\mathbf{f}} \text{ referred to by } \varsigma_j \text{ "corresponds" to the link flow } \tilde{\mathbf{v}} \\ & \text{referred to by } \psi_i \text{, in the sense that } \tilde{\mathbf{v}} = \Delta\,\tilde{\mathbf{f}} \\ 0 & \text{otherwise} \end{cases}$$

The following consistency (equilibrium) condition on the distribution ψ may then be established (Watling, 1999).

Asymptotic equilibrium. Let $\mathbf{t}(\tau^{-1}\tilde{\mathbf{v}})$ denote the cost-flow functions of a network problem, bounded for $\tilde{\mathbf{v}} \in \tilde{\Omega}_2$, and suppose that drivers' conditional route choices are made according to (3.1). Suppose further that drivers form estimates of actual costs from the mean of a random sample $\{\mathbf{T}^{(1)}, \mathbf{T}^{(2)}, \ldots, \mathbf{T}^{(m)}\}$ of the link costs from their previous travel experiences, where m is given, $\mathbf{T}^{(j)} = \mathbf{t}(\tau^{-1}\tilde{\mathbf{V}}^{(j)})$ $(j = 1,2,\ldots,m)$, and $\{\tilde{\mathbf{V}}^{(1)}, \tilde{\mathbf{V}}^{(2)}, \ldots, \tilde{\mathbf{V}}^{(m)}\}$ is a sample of i.i.d., demand-feasible link flow vectors. Then as $m \to \infty$, the (unconditional) link flow probability distribution ψ satisfies the equilibrium condition:

$$\psi = \Pi\, \varsigma(\psi) \tag{3.2}$$

where $\varsigma(\psi)$ is a vector of dimension $|\tilde{\Omega}_1|$ with elements the probabilities

$$\Pr(\, \tilde{\mathbf{F}} = \tilde{\mathbf{f}} \;\big|\; \mathbf{Y} = \mathrm{E}_{\tilde{\mathbf{V}}}[\mathbf{t}(\tau^{-1}\tilde{\mathbf{V}})] \text{ where } \tilde{\mathbf{V}} \sim \psi\,) \qquad (\tilde{\mathbf{f}} \in \tilde{\Omega}_1) \tag{3.3}$$

where $\tilde{\mathbf{V}} \sim \psi$ denotes that $\tilde{\mathbf{V}}$ has a given probability distribution ψ, where $\mathrm{E}_{\tilde{\mathbf{V}}}[.]$ denotes the expectation operator with respect to the distribution of $\tilde{\mathbf{V}}$, and where the conditional distribution of $\tilde{\mathbf{F}} \mid \mathbf{Y}$ is given by (3.1).

The result hinges on the fact that as the number of experiences becomes large, the variance in the mean of their travel cost experiences will approach zero, this mean approaching the true (unconditional) expected cost $\mathrm{E}_{\tilde{\mathbf{V}}}[\mathbf{t}(\tau^{-1}\tilde{\mathbf{V}})]$. But then in the limit, as the costs $\mathbf{Y}$ become deterministic, the *un*conditional distribution of $\tilde{\mathbf{F}}$ will, like the conditional distribution (3.1), also be multinomial, ultimately coinciding with the distribution $\tilde{\mathbf{F}} \big| \mathbf{Y} = \mathrm{E}_{\tilde{\mathbf{V}}}[\mathbf{t}(\tau^{-1}\tilde{\mathbf{V}})]$.

In principle, equilibrium condition (3.2) may be solved for any network problem, giving rise to an equilibrium joint probability distribution of network flows $\tilde{\mathbf{V}}$. By application of the link performance functions as a transformation $\mathbf{T} = \mathbf{t}(\tau^{-1}\tilde{\mathbf{V}})$, the equilibrium joint probability distribution of stochastic link travel costs $\mathbf{T}$ may be computed, from which inferences on reliability may be made. In practice, however, these conditions are not very appealing, as the number of feasible integer link flow states (the dimension of ψ) will be extremely large for realistic networks. It would therefore be expected that any attempt to solve directly the fixed point condition (3.2) would lead to a computationally intensive algorithm.

In order to propose a computationally more attractive notion of equilibrium, we shall then consider approximating the equilibrium condition (3.2). This is achieved by making a second order Taylor series approximation to $\mathbf{t}(\mathbf{v})$ in the neighbourhood of $\mathbf{v} = \tau^{-1}\mathrm{E}[\tilde{\mathbf{V}}]$, leading to a model termed a *Generalised Stochastic User Equilibrium of order 2* (Watling, 1999), and denoted GSUE(2):

Definition: GSUE(2) conditions. Suppose that the cost-flow performance functions $\mathbf{t}(\mathbf{v})$ are twice differentiable at all $\mathbf{v} \in \Omega_2$, and introduce the mean A-vector of flow rates μ and $A \times A$ covariance matrix Σ. Then (μ, Σ) is a GSUE(2) if and only if the following fixed point conditions are satisfied:

$$\begin{cases} \boldsymbol{\mu} = \Delta \,.\, \mathrm{diag}(\boldsymbol{\Gamma}\mathbf{q}) \,.\, \mathbf{p}(\Delta^{\perp}\breve{\mathbf{t}}(\boldsymbol{\mu},\boldsymbol{\Sigma})) \\ \boldsymbol{\Sigma} = \tau^{-1} \; \Delta \,.\, \boldsymbol{\Psi}(\mathbf{q}, \mathbf{p}(\Delta^{\perp}\breve{\mathbf{t}}(\boldsymbol{\mu},\boldsymbol{\Sigma}))) \,.\, \Delta^{\perp} \end{cases} \tag{3.4}$$

where $\breve{\mathbf{t}}(\boldsymbol{\mu},\boldsymbol{\Sigma})$ is an A-vector with elements

$$\breve{t}_a(\boldsymbol{\mu},\boldsymbol{\Sigma}) = t_a(\boldsymbol{\mu}) + \tfrac{1}{2}\left\langle \mathbf{H}_a(\boldsymbol{\mu}),\, \boldsymbol{\Sigma} \right\rangle \qquad (a = 1,2,\ldots, A) \tag{3.5}$$

where $\mathbf{H}_a(\boldsymbol{\mu})$ is the $A \times A$ Hessian matrix of $t_a(\cdot)$ evaluated at $\boldsymbol{\mu}$ (for $a = 1,2,\ldots, A$), where the scalar product of any two n-square matrices $\mathbf{X}$ and $\mathbf{Y}$ is denoted by

$$\langle \mathbf{X}, \mathbf{Y} \rangle = \sum_{i=1}^{n}\sum_{j=1}^{n} X_{ij} Y_{ij}$$

and where $\boldsymbol{\Psi}(\mathbf{q},\mathbf{p})$ is a function whose result is an $N \times N$ block diagonal matrix, with blocks the matrices of dimension $N_k \times N_k$:

$$\Psi_{[k]}(q_k, \mathbf{p}_{[k]}) = q_k \left(\mathrm{diag}(\mathbf{p}_{[k]}) - \mathbf{p}_{[k]} \mathbf{p}_{[k]}{}^{\perp} \right) \qquad (k = 1,2,\ldots,W). \tag{3.6}$$

In the above definition, (3.5) is the second order approximation to expected costs, and $\Psi_{[k]}(\tilde{q}_k, \mathbf{p}_{[k]})$ (note $\tilde{q}_k$, not q_k as in (3.6)) is the covariance matrix of a Multinomial $(\tilde{q}_k, \mathbf{p}_{[k]})$ variable. Once a GSUE(2) solution has been obtained, we may estimate the *variance* and *covariances* in travel times/costs by a similar round of approximation to that used in (3.5). In particular, since

$$\mathrm{cov}(t_a(\mathbf{V}), t_b(\mathbf{V})) = \mathrm{E}[t_a(\mathbf{V}) t_b(\mathbf{V})] - \mathrm{E}[t_a(\mathbf{V})]\mathrm{E}[t_b(\mathbf{V})] \tag{3.7}$$

then clearly $\mathrm{E}[t_a(\mathbf{V})]$ and $\mathrm{E}[t_b(\mathbf{V})]$ may be approximately related to $(\boldsymbol{\mu},\boldsymbol{\Sigma})$ by (3.5), with

$$\mathrm{E}[t_a(\mathbf{V}) t_b(\mathbf{V})] \approx t_a(\boldsymbol{\mu}) t_b(\boldsymbol{\mu}) + \tfrac{1}{2}\langle \mathbf{H}_{ab}(\boldsymbol{\mu}), \boldsymbol{\Sigma} \rangle \qquad (a = 1,2,\ldots, A) \tag{3.8}$$

where $\mathbf{H}_{ab}(.)$ is the Hessian of $t_a(.)t_b(.)$. Note that the covariances in travel times between different links will in general be non-zero, due to the fact that the flows are correlated between links. The link flows are correlated for two reasons: partly due to the fact that the distribution of links flows is built up from the distribution of route flows, and partly, in the case of non-separable cost functions, due to junction flow interactions.

3.2 Stochastic demand

The GSUE(2) model presented above is relatively easily adapted to represent, additionally, variability in the inter-zonal demands. This is achieved by replacing $\tilde{\mathbf{q}}$ by $\tilde{\mathbf{q}}^{\max}$, the *potential* number of travellers on each inter-zonal movement, and for each movement k introducing a dummy "no-travel" path joining each inter-zonal pair, which each potential traveller is assumed independently to choose with a constant probability ε_k. The deterministic-demand GSUE(2) conditions (3.4) are then applied to this extended network, with the result that a stochastic-demand GSUE(2) is estimated for the original (unextended) network. Further details of this approach, with solution methods and numerical examples, are contained in Watling (1998).

3.3 Stochastic capacities

Suppose that for each link $a = 1,2,\ldots,A$, the link performance function is re-denoted $t_a(\mathbf{v},\mathbf{s})$, where $\mathbf{s}$ is an A-vector of given link capacities, in units of vehicles/hour. Then the same asymptotic argument used in section 3.1 leads us to equilibrium conditions of (3.2)/(3.3), except with the expectation $\mathrm{E}_{\tilde{\mathbf{V}}}[\mathbf{t}(\tau^{-1}\tilde{\mathbf{V}})]$ replaced by $\mathrm{E}_{(\tilde{\mathbf{V}},\mathbf{S})}[\mathbf{t}(\tau^{-1}\tilde{\mathbf{V}},\mathbf{S})]$, where the latter expectation is made with respect to the joint distribution of $(\tilde{\mathbf{V}},\mathbf{S})$ where $\mathbf{S}$ denotes the vector of stochastic link capacities.

If $\tilde{\mathbf{V}}$ and $\mathbf{S}$ are statistically independent—as seems reasonable to assume if drivers are unable to anticipate the capacity variations—then

$$\mathrm{E}_{(\tilde{\mathbf{V}},\mathbf{S})}[\mathbf{t}(\tau^{-1}\tilde{\mathbf{V}},\mathbf{S})] = \mathrm{E}_{\tilde{\mathbf{V}}}[\vec{\mathbf{t}}(\tau^{-1}\tilde{\mathbf{V}})] \quad \text{where} \quad \vec{\mathbf{t}}(\mathbf{v}) \equiv \mathrm{E}_{\mathbf{S}}[\mathbf{t}(\mathbf{v},\mathbf{S})] \tag{3.9}$$

and so the only input modification then necessary to (3.3) is to replace $\mathbf{t}(\mathbf{v})$ with $\vec{\mathbf{t}}(\mathbf{v})$ given by (3.9).

Hence, a GSUE(2) problem with stochastic capacities and cost functions $\mathbf{t}(\mathbf{v})$ may be estimated by solving a standard GSUE(2) problem (3.4) but with modified cost functions $\vec{\mathbf{t}}(\mathbf{v})$. This can be used to estimate GSUE(2) flow means, variances and covariances, though the final step (3.7)/(3.8) of estimating *travel cost* variances and covariances needs some modification. In particular, let us assume that $t_a(\mathbf{v},\mathbf{s})$ in fact only depends on s_a (for $a = 1,2,\ldots,A$, so by a slight abuse of notation we can write $t_a(\mathbf{v},s_a)$), and that the stochastic capacities are statistically independent between links. Now consider two distinct links $a \neq b$. Then

$$\begin{aligned}\mathrm{cov}_{(\mathbf{V},S_a,S_b)}(t_a(\mathbf{V},S_a),t_b(\mathbf{V},S_b)) = {} & \mathrm{E}_{(\mathbf{V},S_a,S_b)}[t_a(\mathbf{V},S_a)t_b(\mathbf{V},S_b)] \\ & -\mathrm{E}_{(\mathbf{V},S_a)}[t_a(\mathbf{V},S_a)]\,\mathrm{E}_{(\mathbf{V},S_b)}[t_b(\mathbf{V},S_b)]\end{aligned}$$

and we have, in view of the independence of $\mathbf{V}$ from $\mathbf{S}$, and S_a from S_b:

$$\mathrm{E}_{(\mathbf{V},S_a,S_b)}[t_a(\mathbf{V},S_a)t_b(\mathbf{V},S_a)] = \mathrm{E}_{\mathbf{V}}[\mathrm{E}_{S_a}[t_a(\mathbf{V},S_a)]\mathrm{E}_{S_b}[t_b(\mathbf{V},S_b)]] = \mathrm{E}_{\mathbf{V}}[\vec{t}_a(\mathbf{V})\vec{t}_b(\mathbf{V})]$$

and thus for $a \neq b$,

$$\mathrm{cov}_{(\mathbf{V},S_a,S_b)}(t_a(\mathbf{V},S_a),t_b(\mathbf{V},S_b)) = \mathrm{cov}_{\mathbf{V}}(\vec{t}_a(\mathbf{V}),\vec{t}_b(\mathbf{V}))$$

and in this case we may use precisely the same expressions (3.7)/(3.8) with $\vec{\mathbf{t}}(\mathbf{v})$ in place of $\mathbf{t}(\mathbf{v})$.

The case $a = b$ is a little different, however, since then by the independence of $\mathbf{V}$ and S_a:

$$\mathrm{E}_{(\mathbf{V},S_a)}[(t_a(\mathbf{V},S_a))^2] = \mathrm{E}_{\mathbf{V}}[\mathrm{E}_{S_a}[(t_a(\mathbf{V},S_a))^2]]$$

and so it is necessary to compute additionally the inner expectation before the approximation (3.8) is applied.

4. EXAMPLE 1

As an illustration of the GSUE(2) equilibrium conditions, several simplifying assumptions are made in this section. These are not necessary assumptions, but are only made for the purposes of an example. In particular, we consider the case of a single inter-zonal movement serving a demand rate of q (subscripts will be dropped where possible). Suppose that the movement is connected by A parallel links/routes, numbered in such a way that link a relates to route a (a=1,2,…,A), so that the incidence matrix Δ is the $A \times A$ identity matrix. Suppose that the link cost-flow performance functions $\mathbf{t}(\mathbf{v})$ are 'separable' in the sense that $\frac{\partial t_a}{\partial v_b} = 0 \;\; (a \neq b)$ at all $\mathbf{v} \in \Omega_2$.

Now, dropping the k subscripts, from (3.6) we have:

$$\Psi(q,\mathbf{p}) = q\left(\text{diag}(\mathbf{p}) - \mathbf{p}\mathbf{p}^{\perp}\right) = q\begin{pmatrix} p_1(1-p_1) & -p_1p_2 & \cdots \\ -p_1p_2 & p_2(1-p_2) & \cdots \\ \vdots & \vdots & \ddots \end{pmatrix}. \tag{4.1}$$

(Recall that $\Psi(q,\mathbf{p})$ is not quite an expression for a multinomial covariance matrix, since we are using the demand rate $q = \tilde{q}\tau^{-1}$ rather than the absolute demand $\tilde{q}$.) On the other hand, considering (3.5), we note that in view of the assumed separability of cost functions, the Hessian of $t_a(\mathbf{v})$ will have only one non-zero element: $\frac{\partial^2 t_a}{\partial v_a^{\,2}} \equiv t_a''(\mathbf{v})$, located at the ath diagonal entry of the Hessian. If the corresponding a^{th} diagonal entry in Σ is denoted ϕ_a, then (3.5) yields:

$$\breve{t}_a(\mu,\Sigma) = t_a(\mu) + \frac{\phi_a}{2}\, t_a''(\mu) \qquad \equiv t_a^{\otimes}(\mu,\phi)\text{, say} \qquad (a = 1,2,\ldots,A) \tag{4.2}$$

where the A-vector ϕ denotes the diagonal entries of Σ, namely the link flow variances. Now, (4.2) implies that Σ only enters into the right hand side of the GSUE(2) conditions (3.4) through its diagonal entries ϕ, and so in this case these are the only elements of Σ that are *active* in the equilibration process. This is not to say that the off-diagonal, inactive elements are zero, as can be seen from (4.1); rather, the point is that these off-diagonal terms, the link flow covariances, may be exogenously determined through (4.1) once the GSUE(2) equilibration process has completed.

With these simplifications, the GSUE(2) conditions on (μ,ϕ) become:

$$\mu = q\,\mathbf{p}(\mathbf{t}^{\otimes}(\mu,\phi))\,; \qquad \phi_a = q\tau^{-1}\, p_a(\mathbf{t}^{\otimes}(\mu,\phi))\,(1 - p_a(\mathbf{t}^{\otimes}(\mu,\phi)) \quad (a = 1,2,\ldots,A)\,. \tag{4.3}$$

Making an extension to this basic model to allow for stochastic demand, as in section 3.2, then in (4.3) we replace q by $q^{\max}$ (the potential demand rate), and $\mathbf{p}$ by the joint probability of travelling *and* choosing each of the alternative routes, namely $\varepsilon\mathbf{p}$ where ε denotes the probability of travelling. In fact, since under Binomial demand variation the mean demand rate is $\varepsilon q^{\max}$, then we substitute $q = \varepsilon q^{\max}$ to obtain:

$$\mu = q\,\mathbf{p}(\mathbf{t}^{\otimes}(\mu,\phi)) \qquad \phi_a = q\tau^{-1}\, p_a(\mathbf{t}^{\otimes}(\mu,\phi))\,(1 - \varepsilon p_a\,(\mathbf{t}^{\otimes}(\mu,\phi)))\;(a = 1,2,\ldots,A)\,. \tag{4.4}$$

Clearly, when $\varepsilon = 1$ we recover the deterministic demand conditions (4.3). At the other extreme, for ε very small (and therefore implicitly, for fixed q, the potential driving population on any one day extremely large), the expression approaches $\phi_a = q\tau^{-1}\, p_a(\mathbf{t}^{\otimes}(\mu,\phi))$, which is simply a description of independent Poisson variation (recall the work of Bell, 1991).

Making a further extension to the case of stochastic capacities, as in section 3.3, is straightforward. For example, if the vector variable **S** of link capacities may take one of n possible values $\mathbf{s}^{(1)}, \mathbf{s}^{(2)}, \ldots, \mathbf{s}^{(n)}$, according to user-specified probabilities, then

$$\breve{\mathbf{t}}(\mathbf{v}) \equiv E_{\mathbf{S}}[\mathbf{t}(\mathbf{v},\mathbf{S})] = \sum_{i=1}^{n} \mathbf{t}(\mathbf{v},\mathbf{s}^{(i)}) \Pr(\mathbf{S} = \mathbf{s}^{(i)})$$

and the GSUE(2) conditions are obtained by replacing $\mathbf{t}(.)$ by $\breve{\mathbf{t}}(.)$ in (4.2) before substitution in (4.3)/ (4.4).

5. Solution Algorithm And Example 2

A heuristic solution algorithm for computing GSUE(2) solutions is suggested by the first of the fixed point conditions in (3.4), which for fixed Σ amounts to an SUE condition (2.2) on μ, based on modified link cost functions (3.5). Having solved this problem, yielding equilibrium route proportions **p**, a revised estimate of Σ may then be computed from the second fixed point condition in (3.4). The SUE sub-problem is solved by the method of successive averages (Sheffi, 1985). Outer iteration n of the algorithm is as follows (Watling, 1999):

Auxiliary solution Solve an SUE sub-problem in μ conditional on $\Sigma = \Sigma^{(n-1)}$:

$$\mu = \Delta \,.\, \mathrm{diag}(\Gamma\mathbf{q}) \,.\, \mathbf{p}(\Delta^{\perp}\breve{\mathbf{t}}(\mu, \Sigma^{(n-1)}))$$

denoting the solution by $\overline{\mu}^{(n)}$, and obtain the corresponding Σ from:

$$\overline{\Sigma}^{(n)} = \tau^{-1}\, \Delta \,.\, \Psi(\mathbf{q}, \mathbf{p}^{(n)}) \,.\, \Delta^{\perp}$$

where $\mathbf{p}^{(n)} = \mathbf{p}\left(\Delta^{\perp}\breve{\mathbf{t}}(\overline{\mu}^{(n)}, \Sigma^{(n-1)})\right)$.

Update estimates Update the GSUE(2) estimates according to:

$$\mu^{(n)} = \mu^{(n-1)} + \frac{1}{n}\left(\overline{\mu}^{(n)} - \mu^{(n-1)}\right)$$

$$\Sigma^{(n)} = \Sigma^{(n-1)} + \frac{1}{n}\left(\overline{\Sigma}^{(n)} - \Sigma^{(n-1)}\right)$$

Numerical experience with this algorithm in a range of networks (Watling, 1999) has confirmed the robustness of the algorithm (to factors such as random number seeds), provided a sufficiently large number of iterations are used to solve each "inner" SUE sub-problem.

We now consider an example application of this technique to a simple network, illustrated below.

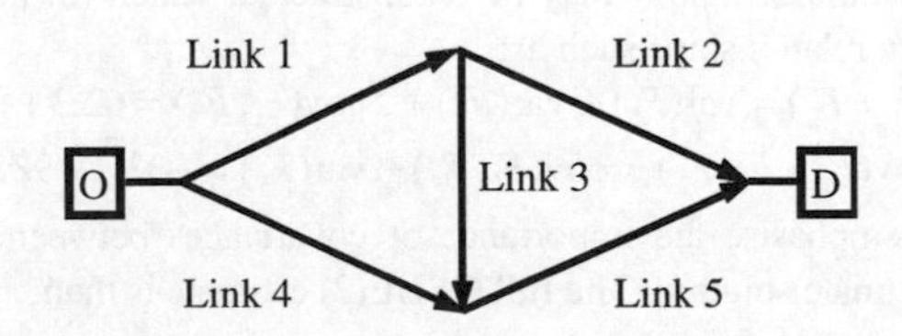

The network has a single inter-zonal movement, serving a demand rate of q=100 vehicles/hour over a time period of duration $\tau = 0.25$ hours, with link cost performance functions of the BPR form:

$$t_1(\mathbf{v}) = 10 + 10\left(\frac{v_1}{30}\right)^2; \quad t_2(\mathbf{v}) = 40 + 20\left(\frac{v_2}{70}\right)^3; \quad t_3(\mathbf{v}) = 5 + 10\left(\frac{v_3}{50}\right)^2;$$

$$t_4(\mathbf{v}) = 50 + 20\left(\frac{v_4}{70}\right)^2; \quad t_5(\mathbf{v}) = 10 + 10\left(\frac{v_5}{30}\right)^2.$$

A probit choice probability function $\mathbf{p}(.)$ is assumed, implicitly defined by assuming independent Normal link cost perceptual errors, $\varepsilon_a \sim \mathrm{Nor}(0, (\alpha t_a(\mathbf{0}))^2)$, where $\alpha > 0$ is a user-specified disperson parameter and $t_a(\mathbf{0})$ is the free flow travel cost on link a $(a = 1,2,...,5)$. A value of $\alpha = 1.0$ was used in the tests below. The solution algorithm was applied with 100 (inner) iterations for each SUE sub-problem, and 20 outer iterations, and stability of the solution was verified by monitoring the between-iteration variation in total travel cost. For comparative purposes, it is noted that the standard SUE solution to this problem was estimated as:

$$\mathbf{v}^{\perp} = (64.0 \quad 46.0 \quad 18.0 \quad 36.0 \quad 54.0) \qquad (\mathbf{t}(\mathbf{v}))^{\perp} = (55.5 \quad 45.7 \quad 6.3 \quad 55.3 \quad 42.4) \ .$$

Since the link cost-flow functions are separable, the only active elements of Σ in the GSUE(2) equilibration process are the link flow variances ϕ (recall the comments made in section 4), and so other elements of Σ are reconstructed after the equilibration is complete. The first test assumed demand and capacities to be deterministic, so that only variability in route choice is represented (as in section 3.1). The GSUE(2) link flow means and variances were estimated as:

$$\mu^{\perp} = (62.1 \quad 42.9 \quad 19.3 \quad 37.8 \quad 57.1) \qquad \phi^{\perp} = (94.1 \quad 98.0 \quad 62.4 \quad 94.1 \quad 98.0) \ .$$

In order to reconstruct the off-diagonal elements of Σ, the GSUE(2) route proportions are required. In fact this step will prove useful as an illustration of the features of the model. In the network defined, there are three routes, and we shall index them as follows:

Route 1 = Link 1 + Link 3 + Link 5
Route 2 = Link 4 + Link 5
Route 3 = Link 1 + Link 2

and the equilibrium route proportions are then easily seen to be $\mathbf{p}^{\perp} = (0.193 \quad 0.378 \quad 0.429)$. Hence the equilibrium route flow rate covariance matrix can be computed as:

$$\mathrm{var}(\mathbf{F}) = \tau^{-1} q \begin{pmatrix} p_1(1-p_1) & -p_1 p_2 & -p_1 p_3 \\ & p_2(1-p_2) & -p_2 p_3 \\ & & p_3(1-p_3) \end{pmatrix} = \begin{pmatrix} 62.3 & -29.2 & -33.1 \\ & 94.1 & -64.9 \\ & & 98.0 \end{pmatrix}.$$

Hence the link flow covariance matrix may be reconstructed, which (in non-matrix notation) is simply using well known relationships such as:

$$\mathrm{var}(V_1) = \mathrm{var}(F_1 + F_3) = \mathrm{var}(F_1) + \mathrm{var}(F_3) + 2\,\mathrm{cov}(F_1, F_3) = 62.3 + 98.0 - 2 \times 33.1 = 94.1$$

$$\mathrm{cov}(V_1, V_2) = \mathrm{cov}(F_1 + F_3, F_3) = \mathrm{cov}(F_1, F_3) + \mathrm{var}(F_3) = -33.1 + 98.0 = 64.9 \quad \text{etc.}$$

This point is made to emphasise the importance of covariances between routes, and therefore between links, in the ultimate solution. The full GSUE(2) estimate is then:

$$\Sigma = \begin{pmatrix} 94.1 & 64.9 & 29.2 & -94.1 & -64.9 \\ & 98.0 & -33.1 & -64.9 & -98.0 \\ & & 62.3 & -29.2 & 33.1 \\ & & & 94.1 & 64.9 \\ & & & & 98.0 \end{pmatrix}$$

Similar computations may be made to yield the travel cost mean and covariance matrix, which we do not present in detail here due to space limitations. The resulting travel cost means and variances are (the covariances exist but are not reported below):

$$E[\mathbf{t}(\mathbf{V})]^{\perp} = (54.0 \quad 45.3 \quad 6.7 \quad 56.2 \quad 47.3) \qquad \mathrm{var}(\mathbf{t}(\mathbf{V}) = diag(176.9 \quad 9.6 \quad 1.4 \quad 9.0 \quad 156.3).$$

Two further tests were performed. The first included stochastic variation in demand, with a probability of travelling of $\varepsilon = 0.8$ assumed. The resulting GSUE(2) mean link flows and variances were:

$$\mu^{\perp} = (62.0 \quad 43.1 \quad 19.0 \quad 38.0 \quad 54.0) \qquad \phi^{\perp} = (125.4 \quad 112.4 \quad 64.0 \quad 106.1 \quad 123.2)$$

with only a minor affect on the means but a significant impact on flow variances, and thereby on travel time variances. The final test included stochastic variation in capacities (the denominator inside the power function of the BPR relationship being assumed to represent a capacity). This was achieved by assuming that link 1 had a capacity of 30 with probability 0.9, and a capacity of 10 with probability 0.1, with suitably modified expected cost functions then supplied to the GSUE(2) equilibration process. With this variation, in addition to the demand variation included in the previous test, the resulting GSUE(2) solution was estimated as:

$$\mu^{\perp} = (53.7 \quad 40.0 \quad 13.8 \quad 46.3 \quad 60.0) \qquad \phi^{\perp} = (123.2 \quad 108.2 \quad 49.0 \quad 116.6 \quad 125.4)$$

with a significant impact on the mean usage of the "risky" link 1.

6. CONCLUSION

A methodology has been presented for estimating both mean network performance and reliability measures within a single consistent framework. Although the model was derived from a fundamental re-definition of the concept of "equilibrium", the resulting model is not so different from conventional approaches; indeed, existing equilibrium computer code is relatively easily extended to accommodate this new model. The work has a potentially significant impact on the transport planning philosophy, as it means that hypothetical policies may be evaluated on a risk basis, not only on their performance in "average conditions". This work would benefit greatly in the future from empirical studies aimed at fitting the model to observed levels of variation, which may in turn lead to the need for more complex models of variability in the input quantities (e.g. demand matrix).

ACKNOWLEDGEMENTS

This research was carried out under the support of an Advanced Fellowship from the UK Engineering and Physical Sciences Research Council.

REFERENCES

Arnott R., De Palma A. & Lindsey R. (1991). Does providing information to drivers reduce traffic congestion? *Transpn Res* **25A**(5), 309-318.

Barcelo J. (1991). Software environments for integrated RTI simulation systems. In *Advanced Telematics in Road Transport*, Proc DRIVE Conf, Brussels, February 1991, Elsevier, Amsterdam, **2**, 1095-1115.

Bell M.G.H. (1991). Expected equilibrium assignment under stochastic demand. Unpublished paper presented at 23rd University Transport Studies Group annual conference, Nottingham, U.K., January 2nd-4th 1991.

Bell M.G.H. and Iida Y. (1997). *Transportation Network Analysis.* John Wiley & Sons, Chichester, UK.

Cantarella G.E. & Cascetta E. (1995). Dynamic Processes and Equilibrium in Transportation Networks: Towards a Unifying Theory. *Transpn Sci* **29**(4), 305-329.

Dial R.B. (1997). Bicriterion traffic assignment: fast algorithms and examples. *Transpn Res* **31B**, 357-359.

Koutsopoulos H.N. & Xu H. (1993). An information discounting routing strategy for advanced traveler information systems. *Transpn Res* **1C**(3), 249-264.

McDonald M. & Lyons G.D. (1996). Driver information. *Traf Engng & Ctl* **37**(1), 10-15.

Mirchandani P. & Soroush H. (1987). Generalized Traffic Equilibrium with Probabilistic Travel Times and Perceptions. *Transpn Sci* **21**(3), 133-152.

Mutale W. (1991). Effect of variability in travel demand and supply on urban network evaluation. Paper presented at 23rd Universities Transport Study Group Conference, January 1991, Nottingham, UK.

Sheffi Y. (1985). *Urban transportation networks.* Prentice-Hall, New Jersey.

Ran B. & Boyce D.E. (1993). *Dynamic Urban Transportation Network Models.* SpringerVerlag, Berlin.

Smith, J.C. & Russam, K. (1989). Some possible effects of AUTOGUIDE on traffic in London. *Preprints, First Int Conf on Vehicle Navigation & Information Systems*, Toronto, Canada, September 1989.

Wakabayashi H. & Iida Y. (1992). Upper and lower bounds of terminal reliability of road networks: an efficient method with Boolean algebra. *Journal of Natural Disaster Science* **14**(1), 29-44.

Watling D.P. (1998). Stochastic Network Equilibrium under Stochastic Demand. Paper presented at 6th EURO Working Group on Transportation, Gothenburg, September 9th-12th 1998.

Watling D.P. (1999). A Second Order Stochastic Network Equilibrium Model. First revision, submitted to *Transpn Sci*.

Willumsen L. & Hounsell N.B. (1994). Simple models of highway reliability—supply effects. Paper presented at *Seventh Int Conf on Travel Behaviour Research*, Santiago, Chile, June 1994.

CHAPTER 4

ILL-PREDICTABILITY OF ROAD TRAFFIC CONGESTION

Serge P. Hoogendoorn and Piet H.L. Bovy, Delft University of Technology, Delft, The Netherlands

4.1 INTRODUCTION

Despite seemingly similar traffic conditions (e.g. traffic flow level, density levels, etc.) it appears that the occurrence of congestion on motorways, expressed in terms of location, day, starting time, ending time, duration and length of queuing, is a highly stochastic phenomenon. At the *macroscopic level*, this is explained to be the result of *unpredictable variations* in travel demand and roadway capacity. Microscopic factors are at work as well, however. It is shown that the complex interaction between drivers at high densities give rise to chaotic-like occurrence of so-called *self-organising localised structures* in the traffic stream. In this chapter, we discuss both the macroscopic level, as well as the microscopic level.

4.2 PREDICTION OF CONGESTION ON MOTORWAYS

Predicting the occurrence and severity of congestion is very similar to weather forecasting. Whether or not it will rain tomorrow in the Netherlands can be predicted to a reasonable level of certainty, but at the level of a particular hour in a particular city this is much more difficult. If a certain level of local and temporal specificity is needed, the prediction is necessarily uncertain and can only be done in terms of probabilities. In the following we will deal with the well-known phenomenon of road traffic congestion.

With regard to traffic flow operations and congestion, Kerner and Rehborn (1996) distinguish three states namely the free-flow state, the transitional state and the congested state. Whereas the *free-flow* state is stable, the *transitional state* is typically *unstable*. That is, small changes in the flow characteristics lead to a more or less sudden shift from a high speed, high flow state into a low speed, low flow state (congestion). Also the reverse may happen. The *congestion state* is characterised by high densities (platooning), low speeds, and stop-and-go waves of smaller or larger platoons of vehicles.

From the perspective of traffic management, which aims at maximising throughput and minimising congestion delays, it is desirable to intervene in the traffic system at the right moment and at the right place in order to prevent the occurrence of congestion or to minimise congestion severity if it cannot be prevented. This would require precise predictions of expected occurrence and severity of congestion. Are such predictions possible? It will be shown in the sequel that congestion is an inherently ill-predictable phenomenon, also in the very short term (minutes or seconds), with characteristics of a chaotic system.

The prediction of congestion in transportation systems is relevant at various planning levels. For the design of traffic management capabilities (e.g. location of variable message signs), medium-term predictions of congestion are necessary (months or year). The daily operations of traffic management (e.g. VMS messages) requires short-term predictions ranging from a minute to a day ahead. Whereas the ill-predictability is apparent at all temporal scales (though in a different way), the sequel will be confined to short-term prediction of congestion (from a few seconds to a few days ahead).

4.3 VARIABILITY IN CONGESTION

If we look at congestion as it manifests itself in the form of traffic queues on the roads, some of the queues are random by their very cause, such as accidents, exceptional weather conditions (black ice) or exceptional traffic conditions (demonstrations, strikes, etc.). From the 16,000 queues on motorways registered in The Netherlands in 1996, about a quarter are unpredictable by their very nature (AVV,1997a). They are thus non-recurrent. The remaining recurrent queues happen from time to time, sometimes almost daily, more or less at the same location. The cause is a temporary lack of roadway capacity, or otherwise stated, a temporary excess traffic demand.

The ill-predictability of recurrent congestion means that the occurrence of congestion in time, place and duration is a highly stochastic phenomenon. Despite seemingly similar traffic conditions (e.g. traffic flow levels), it appears that location, day, starting time, ending time, duration and length of queuing are subject to large and highly stochastic variations. At a number of well-known bottlenecks in the motorway network, congestion is sometimes absent even at working days during rush hours, without any observable cause. On the other hand, phantom jams (also called 'Stau-aus-dem-Nichts') sometimes happen at motorway stretches with sufficient capacity.

Clearly, the ill-predictability differs between locations. At the well-known bottlenecks (bridges, tunnels) or discontinuities (on-ramps) prediction success is higher than elsewhere, but also their precise prediction remains a gambling game with a highly uncertain outcome.

	Starting times of congestion at A20 motorway (during hour no. P.M.)					Queue lengths at A20 motorway (in meters)				
	1992	1993	1994	1995	1996	1992	1993	1994	1995	1996
Mo.	16	--	--	--		4370	--	--	--	0
Tu.	16		--	--	15	3960	0	--	--	3630
We.	17		16	--	15	3000	0	3940	--	2920
Th.	15	16		16		3640	3850	0	3580	0
Fr.	15	16	15	15		5200	3480	2810	3970	0
Mo.		16	15	--		0	3000	5710	--	0
Tu.	15	17	16	16		3000	4000	2000	4670	0
We.	16	16		16		3000	3000	0	3000	0
Th.		15	16	16		0	4210	2000	3380	0
Fr.	16	15	16	16		4770	2980	2610	3690	0
Mo.	16	16	15	16		4000	4190	3410	2860	0
Tu.	16	16		16		3000	3000	0	4000	0
We.		18	15	16	16	0	4000	3520	2570	2740
Th.		17	17	15	16	0	3000	3640	3930	2270
Fr.	15	16	16	15	15	4000	3000	6360	3230	2540
Mo.		16	17	16	16	0	2000	3000	2610	2520
Tu.	16	16		15	16	4000	2270	0	3030	3000
We.	16	16	16	16		4000	3000	2940	3620	0
Th.	16	16	16	16		4000	2000	2270	3400	0
Fr.	16	15	19	15		3000	3000	2400	3280	0
Mo.	16	16			--	3250	2000	0	0	--
Tu.	16	16		15	--	4510	3240	0	3320	--
We.	--	16	16	15	--	--	3300	4610	3900	--
Th.	--	--	16	15	--	--	--	7940	3000	--
Fr.	--	--	--	16	--	--	--	--	2940	--

	frequencies				
	1992	1993	1994	1995	1996
Zero	5	2	6	1	13
15	4	3	4	8	3
16	12	14	9	12	4
17+	1	3	3	0	0
Tot.	22	22	22	21	20

Table 1: Starting time of congestion and queues lengths on A20 motorway downstream of Crooswijk on-ramp, June 1992-1996 (--: no observation).

Using figures from official queue measurements (AVV,1997b), it can easily be shown that congestion is highly stochastic, exhibiting large variations in time and space. Let us for example take the severe recurrent queue location at the A20 motorway after the Crooswijk on-ramp on the Rotterdam beltway, and let us look at the working days of the month of June (1992-1996) in the afternoon (Table 1). On average, on one workday per week there is no congestion at all, the no-queue day varying randomly over the working days of the week. On the days with an afternoon queue, the starting hour varies between 15 p.m. and 18 p.m. with about half of the days' congestion starting between 16 and 17 p.m. On the other days mostly it starts earlier. Also the variation in queue length (if any queue is present) is considerable, between 2 kilometres and 5 kilometres, with a few days having even longer queues up to 8 kilometres.

Similar high degrees of variability in queue characteristics are apparent in the other manifest queue locations of the motorway network in The Netherlands (see AVV (1997a)).

4.4 MACROSCOPIC EXPLANATIONS OF CONGESTION VARIABILITY

We may explain the variability in congestion at various levels. Let us first consider the macroscopic level of trip making. At this level, congestion is a result of a temporary mismatch of demand for and supply of capacity. Both the levels of traffic demand and capacity supply appear to vary randomly to a certain extent. In addition to known systematic variations due to hour, day, and month factors, the level of traffic demand on heavy loaded motorways during peak hours is subject to random variation of at least 5% relative standard variation (see Bexelius (1995)). The *randomness* stems from variability in activity patterns and related travel choices (trip making, origin and destination, departure time, route, mode choices) of road users. The *variability* in daily travel choices is also partly a result of congestion experiences of road users on preceding days (see Figure 1).

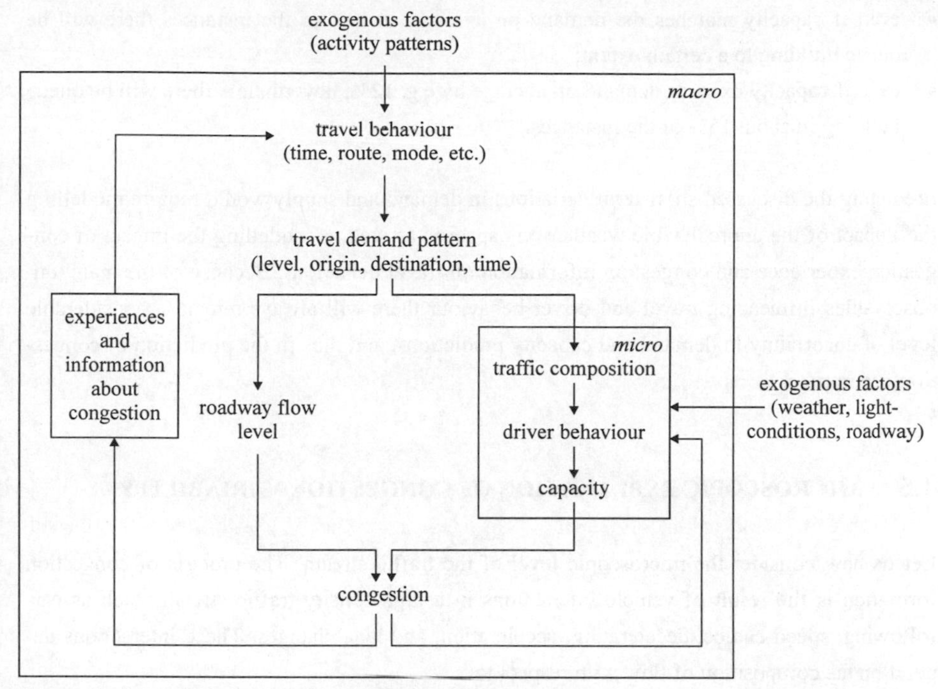

Figure 1: Macro and micro level factors affecting congestion.

It is stressed that the composition of traffic (in terms of mix and sequence of vehicle types, driver types, etc.) at any moment varies even more from day to day. At a given hour of the day, even in the peak, only half of the travellers using a given motorway stretch used that stretch at that hour the day before. This composition variability has a strong influence on capacity as well as on congestion behaviour of a traffic stream. Thus, also capacity is a random variate, the 15-minute value of which for motorways varies with a relative standard variation of about 6% (Van Toorenburg (1986)). Apart from endogenous random factors (such as composition of flow), also exogenous factors are at stake such as weather and light conditions. These levels of variability (estimated at about 10%) do not look excessive. However, in the case of queue building it is the difference between traffic demand and capacity supply that is of importance.

The simultaneous stochastic variation in demand and capacity results in sharp fluctuations in queue formation. The variability in capacity shortage equals roughly speaking about √2 times the average variability in demand and supply, thus about 12% in relative standard deviation. Considering both demand and capacity as independent random variates, this variability implies the following:

- even if capacity matches the demand on average, in half of the instances there will be queue building to a certain extent;
- even if capacity exceeds demand on average by e.g. 12%, nevertheless there will be queue building in about 15% of the instances.

Predicting the discussed short-term variations in demand and supply would require modelling the impact of the unpredictable weather on capacity, as well as modelling the impact of congestion experience and congestion information on travel behaviour. Because of the many unobservables influencing travel and driver behaviour there will always remain a considerable level of uncertainty in demand and capacity predictions, and thus in the prediction of congestion occurrence.

4.5 MICROSCOPIC EXPLANATION OF CONGESTION VARIABILITY

Let us now consider the microscopic level of the traffic stream. The process of congestion formation is the result of vehicle interactions in a high-density traffic stream, such as car-following, speed choice, deceleration, acceleration, and lane changes. These interactions depend on the composition of flow with respect to:

- sequence and mix of *vehicle types* (cars, trucks, busses, vans)
- sequence and mix of *driver types* (risk prone, risk averse, etc.)
- sequence and mix of drivers according to their *information level* (congestion expectancy, familiarity with local conditions, etc.)
- sequence and mix of drivers according to their *trip purpose* (commuters, business-related traffic, recreational traffic, etc.)
- sequence and mix of *driver moods*.

Whether a flow can keep a high level of throughput at a high speed during a longer time without turning into the congestion state depends on all these factors. Only a few of these variables can be known. So, adequate modelling of the occurrence of congestion and its characteristics (time, location, and duration) is only possible to a limited extent. Indeed, various modelling approaches (such as with cellular automata (Ponzlet and Wagner,1996) and fluid-dynamic models (Kerner and Konhäuser,1996)) show that the transition from the unstable to the congestion state is a *spontaneous random process*. In addition, it has been shown that a high-density traffiç flow is a chaotic system (Liu,1991) while Disbro and Frame (1989) have shown that the well-known General Motors car-following model represents a chaotic system if suitable parameter values are chosen. Because we cannot observe the exact starting conditions of this system the prediction of the occurrence of congestion is inherently uncertain.

4.6 Self-Organising Localised Structures in Traffic Flow

The kinetic theory considers the traffic stream as a one-dimensional compressible flow of interacting particles. Although the aim of each driver is to keep moving, driving to one's individual intention becomes gradually impossible when the number of vehicles on the road increases.

It has been found that homogeneous traffic flow in a particular density region can be unstable. Depending on the exact magnitude of the density, the region can be classified as either *stable*, *metastable* or *unstable* (Kerner and Konhäuser,1996).
When traffic flow is operating in the *metastable* or *unstable density area*, traffic flow is characterised by the appearance of complex self-organising structures. We will illustrate the spontaneous appearance of congestion in this metastable state (a so-called *phantom jam*), using a macroscopic traffic flow model.

Recently, a model for the description of multiple vehicle-type traffic flow operations on multilane roadways was developed (see Hoogendoorn (1999)). We consider the lane-aggregate single vehicle-type case. Let $r = r(t,x)$, $v = v(t,x)$, and $q = q(t,x) = rv$ respectively denote traffic density, velocity and flow, at instant t and location x. Moreover, we assume that the velocity variance $\Theta(t,x)$ is constant[1], i.e. $\Theta(t,x) = \Theta_0$. The model of Hoogendoorn (1999) yields the following dynamic equations:

Conservation of vehicles:

$$\frac{\partial r}{\partial t} + \frac{\partial q}{\partial x} = 0 \qquad (1)$$

Flow dynamics:

$$\frac{\partial q}{\partial t} + \overbrace{\frac{\partial}{\partial x}(r\Theta_0 + rv^2)}^{C} - \overbrace{\frac{\partial}{\partial x}\left(\eta \frac{\partial v}{\partial x}\right)}^{T} = \overbrace{r\frac{V(r) - v}{\mathrm{T}}}^{A+D} \qquad (2)$$

The term C of eq. (2) reflects dynamic changes in the flow due to the flux of *energy* $e(t,x) = r(t,x)(\Theta_0 + v(t,x)^2)/2$. That is, the flow q changes due to the balance of inflow and outflow of small groups of vehicles with a different expected velocity v and velocity variance Θ_0. The viscosity term T of eq. (2) delineates the diffusive flux yielding a smoothing of traffic conditions. Kerner and Konhäuser (1996) argue that this term describes the higher-

[1] From traffic observations, we have observed that the velocity variance Θ is a monotonic decreasing function of the traffic density. Although the assumption of a constant variance Θ_0 is not realistic, it is suitable for illustrating the ill-predictability of traffic congestion at the microscopic level that is presented in this chapter.

order anticipation behavior of drivers. Helbing (1997) shows that the viscosity term stems from drivers transitions from brisk to careful driving due to changing traffic conditions.

The right-hand-side term $A+D$ of eq. (2) reflects the combined influence of *acceleration* (A) towards the desired velocity of the free-flowing vehicles, and *deceleration* (D) due to vehicle interactions. That is:

$$\frac{V(r)-v}{\mathrm{T}} = \overbrace{\frac{V_0 - v}{\mathrm{T}}}^{A} - \overbrace{(1-p)\widetilde{P}(t,x)}^{D} \tag{3}$$

Let us briefly discuss the terms in eq. (3). V_0 and T respectively denote the *acceleration velocity*, and the *acceleration time* of the vehicles in the traffic stream. Hoogendoorn (1999) shows that the acceleration velocity V_0 is defined by the *average desired velocity* of the free-flowing (platoon-leading, unconstrained) vehicles in the flow. This holds equally for the acceleration time T, reflecting the time free-flowing vehicles need to accelerate towards V_0.

The *immediate lane-changing probability* $p = p(r,v)$ denotes the probability that a vehicle is able to immediately change lanes, without needing to reduce its velocity. The *modified traffic pressure* $\widetilde{P}(t,x)$ is defined by:

$$\widetilde{P}(t,x) = (1-\phi(t,x))\ell(t,x)P(t,x) \text{ with } P(t,x) = r(t,x)\Theta_0 \tag{4}$$

The modified traffic pressure describes the *expected number of vehicle interactions per unit time per vehicle*. An interaction occurs when a fast vehicle in the stream catches up with a slow vehicle. Moreover, it is implicitly assumed that vehicles unable to pass the impeding vehicle will form a platoon, and consequently become constrained. The *fraction of these constrained vehicles* $0 \le \phi(t,x) \le 1$ is included to describe the correlation between unconstrained (platoon-leading) vehicles, and constrained (platooning) vehicles. The *vehicle-spacing correction factor* $\ell(t,x) \ge 1$ is defined by:

$$\ell = (1 - r(L + vT + (v^2 + \Theta_0)F))^{-1} \tag{5}$$

where L, T, and F respectively denote the average vehicle length, the reaction time, and the speed-risk factor (Jepsen, 1998). The correction factor $\ell(t,x)$ reflects the *increased number of interactions* due to consideration of *vehicle spacing requirements*. For an elaborate discussion of the model, we refer to Hoogendoorn (1999).

The acceleration term (A) and the deceleration term (D) in eq. (3) can be considered as reflecting *competing processes*: on the one hand, drivers aim to accelerate towards the acceleration velocity V_0, which is an so-called *active process* (Kerner and Konhäuser,1996); on the other hand, drivers need to decelerate when catching up with slower vehicles in the stream without having the opportunity to change lanes. This can be considered a *damping process* (Kerner and Konhäuser, 1996).

Let us assume that the immediate lane-changing probability $p(r)$ is specified such that:

$$V(r) = V_0 \left(\frac{1}{1 + \exp((r / r_{jam} - 0.25) / 0.06)} - 3.72 \cdot 10^{-6} \right) \tag{6}$$

where r_{jam} denotes the so-called *jam density*, yielding the model of Kerner and Konhäuser (1996). They use an analogy of traffic system with other, active physical systems to explain observed phenomena in traffic, using specific properties of *localised structures* spontaneously appearing in the traffic flow.

The state of *low-density traffic*, where interactions (reflected by the *modified traffic pressure* $\widetilde{P}(t,x)$) between vehicles are rare, is both *regular* as well as *predictable*. This *traffic state* is stable, in the sense that small perturbations in the flow (for instance, localised changes in density due to a vehicle spontaneously braking) will generally diminish very fast by drivers anticipation (reflected by the viscosity term T of eq. (2)). However, if the density increases, local interactions occur more frequently, whilst their intensity becomes stronger. These interactions cause the *damping process* to lead to the spontaneous formation of localised structures (phantom jams).

4.7 Example of Self-Organisation of Congestion

To illustrate the process of *self-organisation*, let us consider a hypothetical case. Consider a circular road of 20km. Let us assume that the initial density distribution $r(0,x) = r_0(x)$ prescribes slightly perturbed homogeneous traffic conditions:

$$r(0,x) = r_0 + \delta r_0 \cos(2\pi x / L) \tag{7}$$

where we have used $r_0 = 23$veh/km. Let us assume $v(0,x) = V(r_0)$. To show the chaotic-like behaviour of the traffic system, three scenarios are considered that are characterised by slightly different initial conditions, namely: $\delta r_0 = 1.0$, 2.0 and 3.0veh/km respectively. Figure 2 shows the kinetics of local cluster formation for all scenarios.

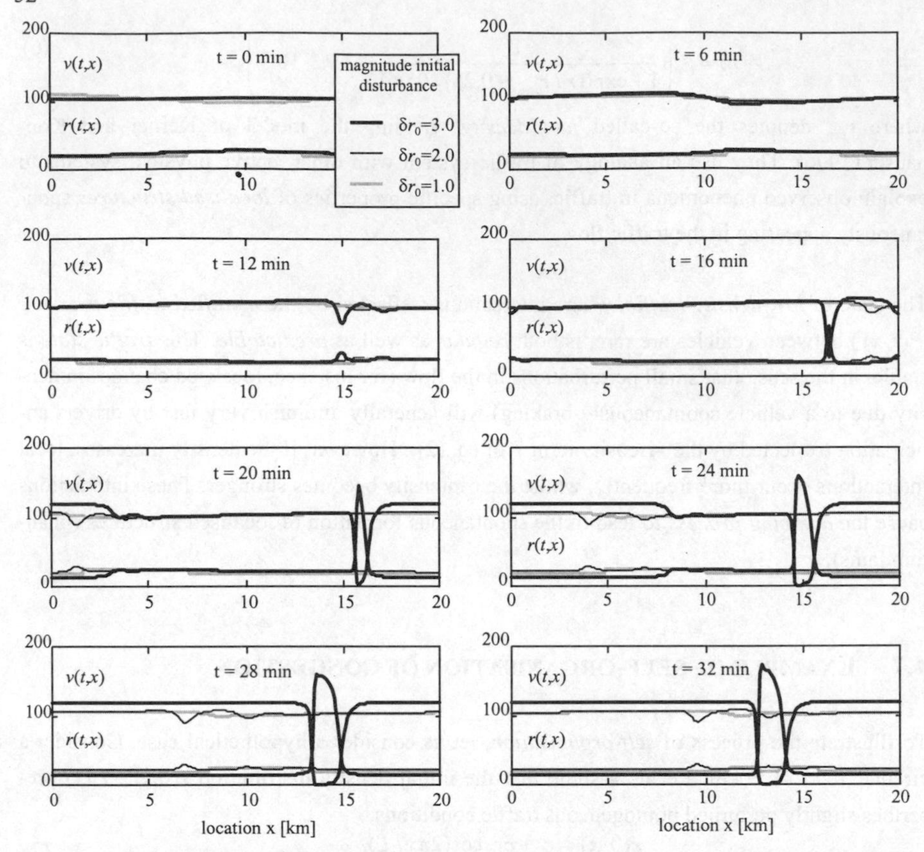

Figure 2: The density and velocity evolution for different density distributions δr_0=1.0, 2.0, and 3.0, using the parameter values: r_{jam} = 175veh/km, V^0 = 120km/hr, T = 6s, $\Theta_0 = 13m^2/s^2$, and η = 60m/s.

When the initial perturbation is very small ($\delta r_0 = 1.0$), no traffic breakdown occurs. Figure 2 clearly shows how drivers anticipate on downstream traffic conditions react accordingly, thereby preventing a localised structure to appear.

In the other cases ($\delta r_0 = 2.0, 3.0$) however, the perturbations *do* result in a local breakdown of traffic, although neither appears at *the same time*, nor at *the same location*. In these cases, the perturbation is critical (Kerner and Konhäuser,1996). At first, this critical perturbation has a small amplitude which increases very slowly in time. Gradually, the shape of the critical perturbation is deformed, and after some time, a local high amplitude perturbation of the density appears. When this traffic breakdown has occurred, vehicles arriving at a higher density region are held back, resulting in the formation of a low density region downstream of the traffic jam. Further downstream, a *transition layer* is formed between the low density region and the slightly disturbed homogeneous flow. In this transition layer, fluctuations are present.

These oscillations have the same character as the local perturbation leading to the local cluster. In this particular case, the oscillations gradually disappear. However, it is conceivable that under slightly different circumstances, the oscillations may again lead to a local breakdown of traffic.

This example showed the formation of a *self-organised structure*, governed by a *local critical* disturbance. In this respect, both the magnitude of r_0 and δr_0 of the perturbation are of dominant importance: they determine the *time* and *location* of the formation of the phantom jams in a *chaotic* manner. That is, the experiment shows that very small differences in these initial perturbations of the initial freeway conditions result in very large differences with respect to the starting time and location of the traffic jam, or whether congestion occurs at all. Note that Kerner and Rehborn (1997) also provide empirical evidence regarding this non-linear theory of traffic clusters.

4.8 CONCLUSION

Summarising, the process of local traffic breakdown of traffic in a particular density region is a process of self-organisation and snowball-like growth of local perturbations. That is, in the presence of (possibly small) non-homogeneities caused by on-ramps, off-ramps, bottlenecks, lane changes, slowly moving vehicles, etc., traffic jams can be formed, even when the *initial fluctuations are negligible*. The exact magnitude of these perturbations are uncertain and unpredictable due to the stochastic nature of the traffic flow's driver- and vehicle-composition, the traffic demands at the on- and off-ramps, etc. This implies that the determination of the exact time and location of traffic jams traffic conditions is impossible, because of the chaotic-like behaviour of traffic flow under metastable or unstable traffic conditions.

Congestion, thus, appears to be an ill-predictable phenomenon. Because the traffic pattern is a social process with complex interactions between travellers, we cannot fully understand the mechanisms, nor can we include in the models the many unobservables that play a role. In addition, it has been shown that the traffic process at high densities is a chaotic system the outcome of which strongly depends on the initial conditions. Because these initial conditions cannot be determined exactly, precise predictions of future states are inherently limited.

4.9 REFERENCES

AVV (1997a), *Yearly Traffic Data Report 1996* (in Dutch). Rotterdam, Ministry of Transport, 1997.

AVV (1997b), *Extract From Queue Data Files*. Rotterdam, Ministry of Transport, 1997.

Bexelius, S. (1995), *Fluctuations In Traffic Intensities. How Normal Is Normal*? Verkeerskundige Werkdagen, CROW (in Dutch).

Disbro, J.E. and M. Frame (1989), *Traffic flow theory and chaotic behaviour*. Transportation Research Record No.1225, 109-115.

Hoogendoorn, S.P. and P.H.L. Bovy (1997), *A Macroscopic Multiple User-Class Traffic Flow Model*. In: Proceedings of the 3rd TRAIL PhD Congress, The Hague.

Kerner, B.S., P. Konhäuser, and M. Schilke (1996), *A New Approach to Problems of Traffic Flow Theory*. In: Proceedings of the 13th International Symposium of Transportation and Traffic Theory, INRETS, Lyon.

Kerner, B.S., and H. Rehborn (1996). *Experimental Features and Characteristics of Traffic Jams*. In: Physical Reviews E 53(2), 1297-1300.

Liu, G.Q. (1991), *Congestion, Flow And Capacity*. In: Brannolte,U.(ed) Highway capacity and level of service. Rotterdam, Balkema, 245-251.

Ponzlet, M. and P.Wagner (1996), *Validation Of A CA-Model For Traffic Simulation Of The North-Rhine-Westfalia Motorway Network*. PTRC, Proceedings P404.

Van Toorenburg, J.A.C. (1986), *Practical Capacity Values* (in Dutch). Verkeerskunde 37(5/6).

Jepsen, M. (1998), *On the Speed-Flow Relationship in Road Traffic: A Model of Driver Behaviour*. Proceedings of the Third International Symposium on Highway Capacity, 297-319.

CHAPTER 5

TRAFFIC CONTROL AND ROUTE CHOICE: CONDITIONS FOR INSTABILITY

Henk J. van Zuylen and Henk Taale, Delft University of Technology, Department of Civil Engineering and Geosciences and Transport Research Centre of the Ministry of Transport, the Netherlands.

INTRODUCTION

Traffic control and traveller's behaviour are two processes which influence each other. The two processes have different objectives which the actors try to achieve and the decisions taken in traffic control have influence on the possibilities for travellers to choose their preferred mode, route and time of departure, and vice versa. The result of the optimisation of the different actors on the dynamics of the system and the existence of stable situations where each actor has no possibility to optimise his decisions any more have been studied by simulation and graphical analysis.

Under certain conditions multiple stable situations are possible, but some of these situations are sensitive to small disturbances, by which the system moves away from the original situation. There appears to be a non-linear relationship between system parameters and the character and location of the equilibrium situations.

THE RELATIONSHIP BETWEEN TRAFFIC CONTROL AND ROUTE CHOICE

There is a tradition in traffic control to adapt control structure and parameters to the actual traffic flows and conditions, such that delays, queues or some other collective objective function is optimised. For fixed time and traffic actuated controllers, methods and guidelines have been developed, such as Webster's formula for cycle time and green splits, the rule to minimise delays by choosing green splits which minimise the maximum degree of saturation, the rules for maximum green times and maximum gap times for vehicle actuated controllers, etc.

Apart from the optimisation of the total network, often a special treatment is given to selected groups of road users. In urban areas pedestrians may get a preferred treatment, at bus routes the bus may get a priority treatment and on crossings of cycle tracks cyclist may get more frequently a green phase. Such priority control is the consequence of a policy to assign the use of road space especially to certain preferred groups of road users because they play an important role in the local situation. In the case of priority for public transport, a reason might be that this gives a reduction of operating cost for a transport mode which is already heavily subsidised by public funding.

There is also a certain expectation from the authorities that a preferential treatment of certain traffic classes - and in most cases one has public transport and pedestrians in mind - will have the consequence that this will reduce the growth of car traffic and will influence the modal choice in favour of collective public transport, cycling or walking. In each case it is possible to influence route choice and time of departure by traffic control.
Because traffic control has an influence on travel behaviour, a change in traffic control may have an impact on traffic volumes. If traffic control is modified such that congestion on a certain route disappears and delays on intersections decrease, traffic might be attracted from other areas where congestion still exists or which are part of a longer route. This might have the consequence that queues which originally disappeared, return. Delays may come back to the original levels. The question is whether there still is a net profit for the whole traffic system.

If we assume that a modification of the traffic control gives a change in travel behaviour, it is necessary to anticipate this change. If we want to optimise delays, it should be done for the traffic volumes which will be present *after* the introduction of the optimised traffic control and not for the traffic volumes which exist *before* the implementation. Of course, it is possible to follow an iterative approach, where after each shift in traffic volumes the control scheme is adjusted until equilibrium has been reached, or one may use a self-adjusting traffic control. However, it can be shown for certain examples that the process of the adjustment of traffic control followed by a shift in traffic volumes, does not necessarily lead to a system optimum. It is even possible that the system oscillates between two or more states.

In this chapter we start with the description of a simple route choice model which displays the occurrence of oscillations. We will study a simple road network with traffic control on one of the links and we investigate the occurrence of stable, consistent equilibrium situations. First we give the result of a simulation study. This shows the regular pattern that one stable condition exists where flows are consistent with the travel times and the traffic control is optimised with respect to total delay. A further analysis of a slightly simplified situation shows that there is a possibility of the occurrence of oscillation in route choice for a simple road network. Afterwards, we study the combined route choice and traffic control optimisation problem in more detail. It is shown that in the two-level optimisation problem, optimum solutions exist which can be interpreted as *meta-stable saddlepoints*. This means that there is equilibrium between traffic control and route choice, but that small disturbances in the conditions will cause the system to slide to another state.

ROUTE CHOICE BASED ON EXPERIENCE FROM THE DAY BEFORE

The route chosen by travellers depends on their perception of travel times on different alternative routes. Normally, travellers have imperfect knowledge about the actual travel times and have to rely on experiences in the past. It has been shown that route choice based on historical knowledge can lead to oscillating behaviour, where on one day one route is preferred and the next day most of the traffic follows another route (Nakayama et al. 1999).

A simplified model which describes the dynamics of route choice with two alternative parallel routes is the following. We assume that on day n the volume on route 1 is proportional to the volume of the previous day on the same route (conservative force) and also proportional to the volume on the other route (the more volume on the other route, the higher the travel times and the more travellers will switch).

$$V_{n,1} = \alpha V_{n-1,1} V_{n-1,2} = \alpha V_{n-1,1} (V - V_{n-1,1}), \tag{1}$$

where $V = V_1 + V_2$ and α is a constant which can be determined from observations, e.g. from the equilibrium state when $V_n = V_{n-1}$.

Equation (1) is an example of the logistic equation, which is known to give equilibrium states for certain values of α, oscillating behaviour for other (higher) values of α and chaotic behaviour if α comes above a certain value. The dynamics of the system depend, apart from α, also on the initial value of V.

If in equation (1) the transition is made from volumes to fractions of traffic choosing a route, x, the equation becomes

$$x_n = \alpha' x_{n-1} (1 - x_{n-1}) \tag{2}$$

where $\alpha' = \alpha V$.

In Figures 2 and 3 the dynamics of route choice is illustrated for different values of α. The transition between different dynamic patterns can be described as splitting of the number of states between which the route choice oscillates. The transition to chaotic behaviour is characterised by the fact that there is an unlimited number of states between which the route choice moves.

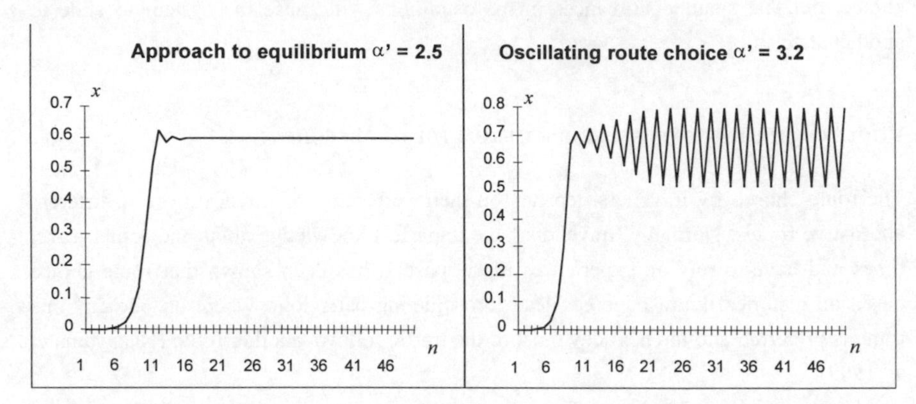

Figure 1 Route choice between two alternative routes, equilibrium and alternating behaviour.

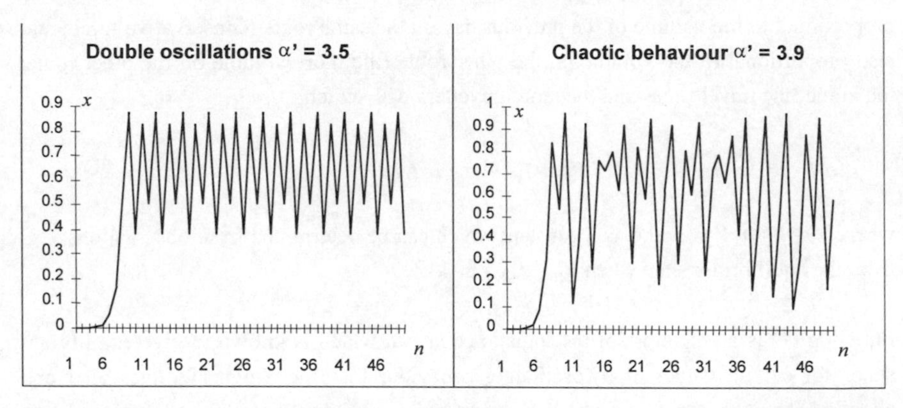

Figure 2 Route choice: oscillations with four states and the transition to chaotic behaviour.

The difference between random and chaotic behaviour is, that chaotic behaviour has certain regularities, such as the quasi-periodicity and (strange) attractors. Furthermore, chaotic behaviour often has the property that small changes in behaviour result in large changes in the future state. The similarity is that both random and chaotic behaviour cannot be predicted. Important properties of systems with chaotic properties are:

- they have non-linear dynamics
- positive feedback exists, by which certain changes are enhanced
- negative feedback exists, which drives the condition of the system from a dynamics of unlimited growth in one direction.

These three properties can clearly be seen in the example in this section: equation 2 is non-linear, for small values of x there is a positive feedback proportional to x and for values of $x \approx 1$ the negative feedback pushes x back in the direction of smaller values.

THE COMBINED TRAFFIC ASSIGNMENT AND TRAFFIC CONTROL OPTIMISATION PROBLEM

Already in 1974, Allsop showed the relevance of the interdependence of traffic control and route choice. The problem to be solved was initially to search for a traffic control scheme that optimises total delay for traffic volumes which are consistent with the travel times influenced by the control scheme, i.e. a traffic condition where no traveller can improve his travel time by choosing another route. Two parties try to achieve their own goals, each with its own objective function and space of choices (Charlesworth 1977, Cascetta et al. 1998). The infrastructure manager tries to optimise the road system, to maximise the utilisation and to minimise total delays and stops. A part of the available instruments are the settings of traffic signals. The road user tries to minimise his own travel time and to reach his destination in time. The choices he has include the routes and the departure time. Fisk (1984) shows that this situation can be seen as an example of a non-co-operative game, in which two players have their own objectives and their own strategies. The strategy is known and the choices are predictable, such that it is possible to choose an optimal strategy, taking into account the predictable reaction of the other party. Road users can choose their route under the assumption that traffic control will be optimised for the total delays. The infrastructure manager can optimise the traffic control knowing that the road users will shift their route choices after the modification of the control scheme.

AN EXAMPLE

A first attempt to analyse this problem was made by a simulation study. The objective was to analyse the distribution of flows between one origin A and one destination B between which two alternative routes are available. Route 1 uses a controlled intersection, route 2 is a bypass.

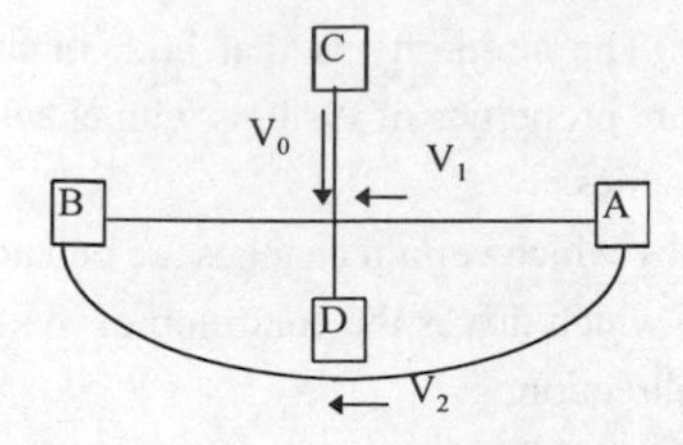

Figure 3 Simple road network to illustrate the coupling between route choice and traffic control.

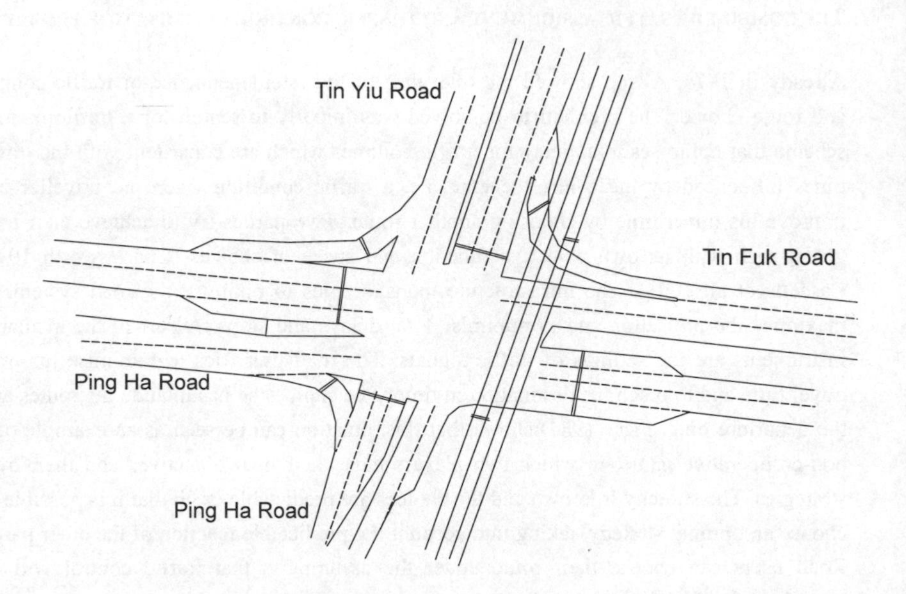

Figure 4 The controlled junction on the route between A and B.

A junction from the New Territories of Hong Kong was chosen as example for this study. The junction is sketched in Figure 4. For this junction a two-lane bypass for the north-south movement was created. This bypass is situated on the West Side of the junction and is 2.5 kilometres longer than the route across the junction. The free-flow speed for the bypass is 100 km/hr and for the route with the junction 50 km/hr. The flows for the AM-peak are as given in Figure 4. These are, apart from a small change, the original flows.

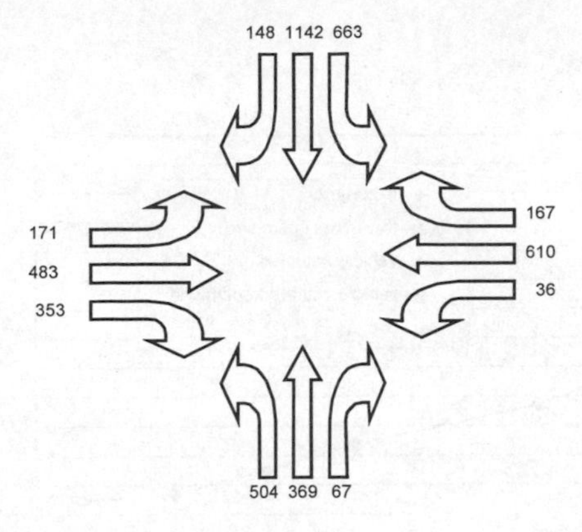

Figure 5 Traffic flows AM peak.

For this situation a number of control strategies were simulated to see how it affected route choice. For the simulation the microscopic simulation model FLEXSYT-II- was used. FLEXSYT-II- has been developed by the Transport Research Centre (AVV) of Rijkswaterstaat. A more detailed description of the model can be found in (Taale and Middelham 1995 and 1997) and (Taale and Scheerder 1998).

Because FLEXSYT-II- has no assignment, other than specified by the user, route choice in this study was investigated by changing the splitting rate. First, it was assumed that only 1% of the traffic took the bypass (to measure travel time), then the splitting rate was increased to 10%, 20%, etc. In this way it was possible to find the equilibrium. This was done for four control types. First, the existing fixed-time control plan was simulated for all splitting rates. Then the same was done for optimised fixed-time control. The optimised control plan was derived by using Webster's formula for cycletime and greentimes for the current flows. Then, the existing vehicle actuated control plan was used and finally an optimised vehicle actuated control. This was done also by using Webster's formula and putting the greentimes as maximum greentimes in the control plan.

In the next figure the travel times on both routes are shown for all control types. The travel time for the bypass is the same for all types.

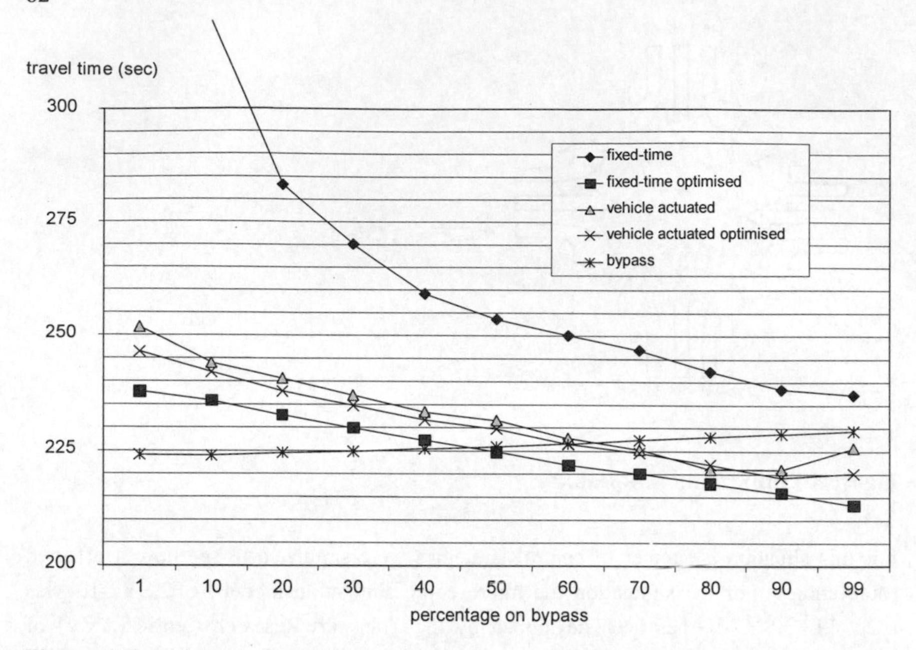

Figure 6 Travel times for different distributions of the flow on the controlled route and the bypass.

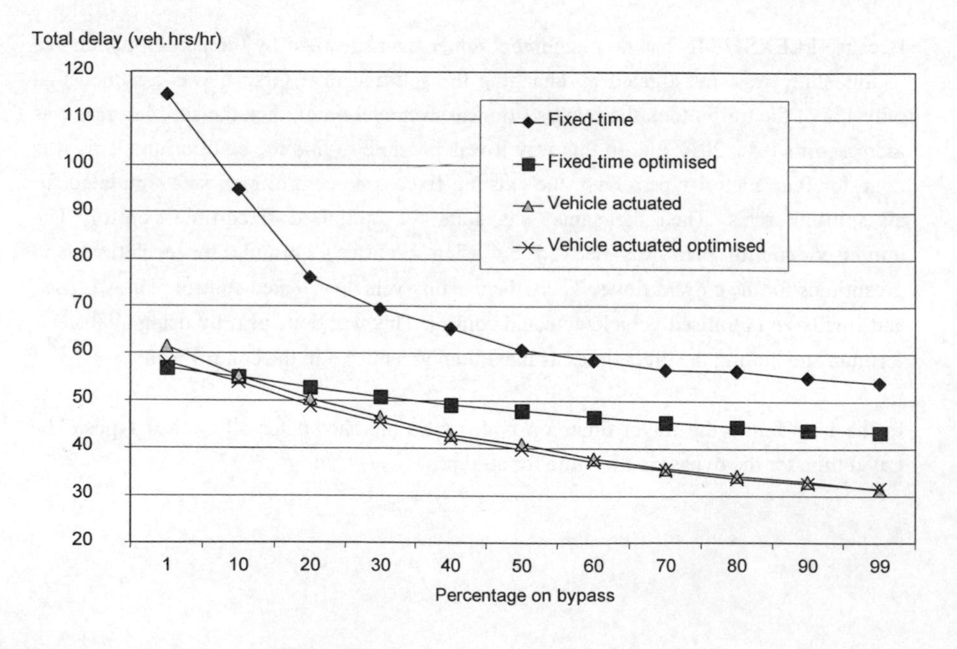

Figure 7 Total delay on both links.

For normal fixed-time control equilibrium is never reached. For optimised fixed-time control the equilibrium is around 50%. For both vehicle actuated control types equilibrium is reached around 60%. The total delay is shown in Figure 7. The figure shows that a user optimum does not necessarily means a system optimum. From the previous figure it was clear that a user optimum was located around 50% or 60%, but from Figure 7 it can be seen that a system optimum is reached when as much traffic as possible uses the bypass.

So far, the green times were optimised only for the situation in which only 1% of the traffic uses the bypass. When the optimisation is done for both equilibrium situations, the following table is the result.

Table 1: Results before and after optimisation at equilibrium.

	Fixed-time (splitting rate 50%)		Vehicle actuated (splitting rate 60%)	
	before	*after*	*before*	*after*
total distance travelled (veh.km/hr)	10888.97	10823.72	11080.76	11080.47
travel time bypass (sec.)	225.9	225.8	226.5	226.5
travel time junction (sec.)	224.8	224.9	226.5	226.3
total delay (veh.hrs/hr)	47.91	39.14	37.25	37.26

Table 1 shows that for fixed-time control the travel time for the two separate routes does not change at all: the total delay decreases by 18%. For vehicle actuated control the situation does not change at all: the travel times on the two routes and the total delay stay the same.

MULTIPLE SOLUTIONS

The result of the simulation study showed one single solution for the combined problem of assignment and traffic control. In order to investigate the possibilities for equilibrium solutions and to look for multiple solutions the network has been changed slightly without changing the degrees of freedom. In Figure 8 the second network layout is given:

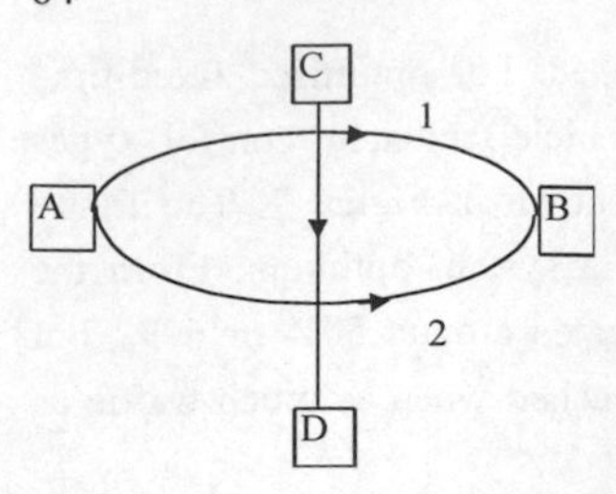

Figure 8 Example of a network with symmetric choice possibilities.

The example of Figure 8 is symmetric. Both routes between A and B cross the route between C and D at a controlled intersection. The traffic control at these intersections is fixed-time optimised with Webster's method. It appears that, depending on the magnitude of the flows and the internal lost time, several different solutions exist:

1. the symmetrical solution with a 50-50 % distribution between route 1 and 2
2. an asymmetrical solution where more drivers choose one of the two routes.

Of course, for every asymmetrical solution a 'mirror' solution exists: if a stable solution is obtained with x % on route 1 and 100 minus x % on route 2, another solution is the distribution 100 minus x % on route 1 and x % on route 2.

In Figure 9 the difference in travel time between routes 1 and 2 is given for different values of the percentage of the traffic using route 1. The stable situations occur if the difference in travel time is zero. Figure 9 gives the symmetrical situation where the equilibrium and optimum is obtained for 50% distribution.

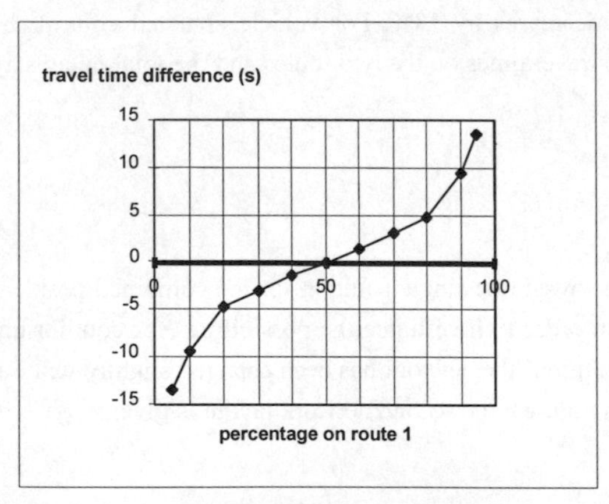

Figure 9 Travel time difference between routes 1 and 2, giving one single symmetric equilibrium.

The situation was calculated for 1000 veh/hr flow from A to B, 500 veh/hr from C to D, saturation flows of 1800 veh/hr and internal lost times of 9 seconds, with a minimum green time of 6 seconds.

If we change the flow from A to B to 600 veh/hr, the picture changes a lot:

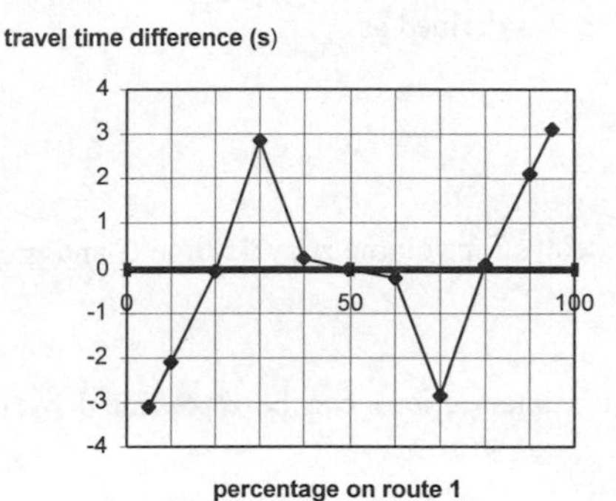

Figure 10 Travel times differences for routes 1 and 2 with asymmetric equilibrium.

Three different states exist, where the travel times for both routes are equal. Two of them are stable: the (20 - 80 %) and (80 – 20%). The (50 - 50%) distribution is also an equilibrium in the sense that travel times are equal for both routes and drivers have no reasons to change their route choice. However, if a change in the distribution occurs, this change is enhanced subsequently by the adaptation of the traffic control scheme to the new traffic flow distribution: the control scheme for the route with the larger flow will get shorter delays. The system optimum, with the minimum total travel time is in both cases (Figures 9 and 10) the 50 - 50% distribution.

Changing the parameters of the control scheme (lost time or minimum times), the flow or the saturation flows changes the appearance of the time-difference curves significantly, so that a small change of the parameters can have the result that the equilibrium states move over large distances and the a-symmetric solution disappears suddenly.

Apparently the system of route choice and traffic control is under certain circumstances critically dependent on system parameters. In the following section we shall investigate this behaviour in more detail for the original network Figure 3.

FURTHER ANALYSIS

The assignment problem in the case of deterministic route choice based on shortest route choice by individual drivers, can be formulated mathematically as

$$\min_{(V)} Z(V_i) \qquad (3)$$

where V_{i+} is the volume on link i. The function Z is defined as

$$Z = \Sigma_i \int^{V_i} T_i\,(z, C, t_g\,)\,\mathrm{d}z \qquad (4)$$

and T_i is the travel time on link i including delays for volume z, cycle time C and green time t_g.

The minimisation of the delays on controlled intersections can be represented by the following formal expression

$$\min_{\{t\}} \Sigma_i\, D_i\,(V_i,\, t_{ij}) \qquad (5)$$

where t_{ij} are the time parameters of the traffic control on link i, D_i is the delay for link i and V_i represents the volumes to be calculated from the solution of eq. (3).

In order to get some more insight into the characteristics of the problem, we shall reduce the traffic control problem to one single dimension. The combined assignment and optimisation problem can be visualised in a two dimensional space which makes further analysis easier. We assume that the cycle time remains fixed, the only parameter left is the green split.

The delay D for a single controlled flow is given by

$$D(V, C, t_g) \approx 0.9\,[\tfrac{1}{2}\,(C - t_g)^2\,(1 - V/s)^{-1}\,C^{-1} + \tfrac{1}{2}\,x^2\,/V\,(1 - x\,)\,] \qquad (6)$$

with

- V = volume
- $x = (V/s)\,(C/t_g)$
- s = saturation flow
- C = cycle time
- t_g = green time.

For both approaches of the intersection together the total delay is given by

$$\Sigma D_i \approx 0.9\,[V_1\,\{\tfrac{1}{2}(C - t_{g1})^2 (1 - V_1/s_1)^{-1} C^{-1} + \tfrac{1}{2}(V_1 C/t_{g1} s_1)^2 / V_1 (1 - V_1 C/t_{g1} s_1)\} + V_0\,\{\tfrac{1}{2}(C - t_{g0})^2 (1 - V_0/s_0)^{-1} C^{-1} + \tfrac{1}{2}(V_0 C/t_{g0} s_0)^2 / V_0 (1 - V_0 C/t_{g0} s_0)\}] \quad (7)$$

with

$0 \le V_1 \le V$

$t_{g1} + t_{g0} = C - t_l$

t_l is the internal lost time of the control scheme..

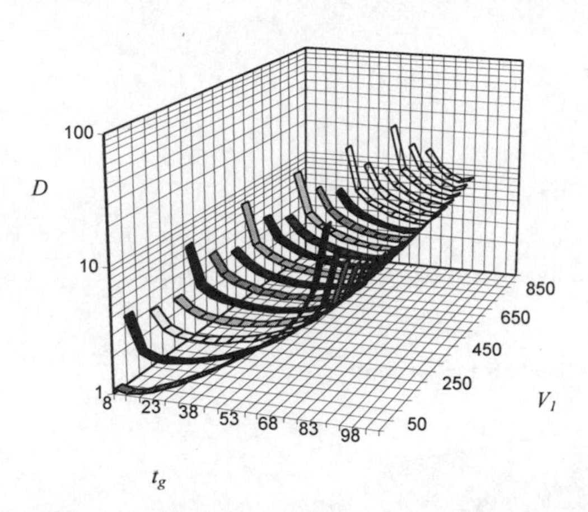

Figure 11 Total delay as a function of the green time t_{g1} and volume V_1.

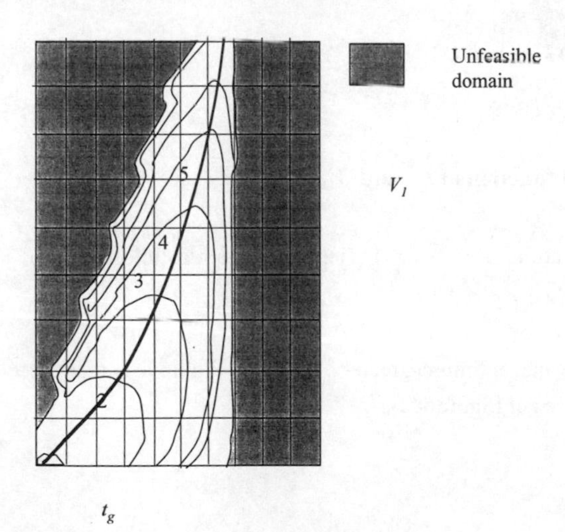

Figure 12 Iso-curves with equal total delay in the t_g - V_1 plane. The thick line gives the green time that minimises the total delay for a given V_1.

The route choice problem can be formulated in the following way

$\min_{\{V\}} Z\{V_1, V_2\}$ with $V_1 + V_2 = V$(8)

where Z can be elaborated to

$$Z = V_1 L_1/v + V_2 L_2/v + 0.45 \left[(C-t_{g1})^2 / C \int (1 - z / s)\, dz + (C / s.t_{g1})^2 \int z (1 - z C / s.t_{g1})\, dz\right]$$
$$= V_1 L_1/v + V_2 L_2/v + 0.45\left[s\, (C-t_{g1})^2 / C \ln (1 - V_1 / s)^{-1} - V_1 C / s.t_{g1} - \ln (1 - V_1 C / s.t_{g1})\right] \qquad (9)$$

where L_i is the length of link i. For eq. (9) the boundary condition $V_1 + V_2 = V$ applies, the function Z becomes also a function of two variables, V_1 and t_{g1}.

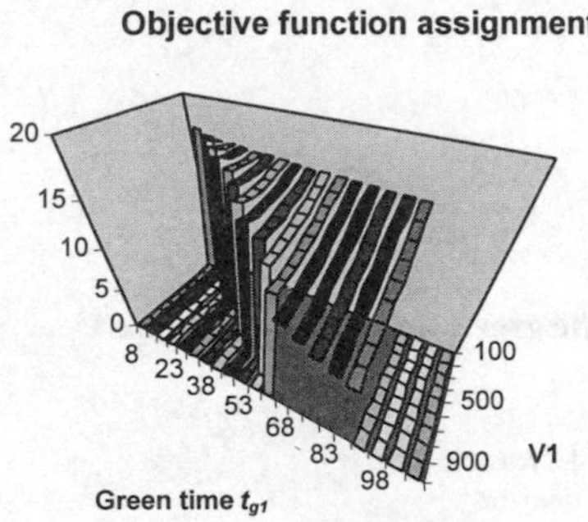

Figure 13 Objective function Z as a function of t_{g1} and V_1.

Figure 13 gives a graphical representation of $Z(V_1, t_g)$. The equilibrium solutions are on the line drawn in Figure 14 where $\delta Z / \delta V_1 = 0$.

If we combine the lines which give the optimum green split (from Figure 12) and the equilibrium assignment (Figure 14) we get Figure 15.

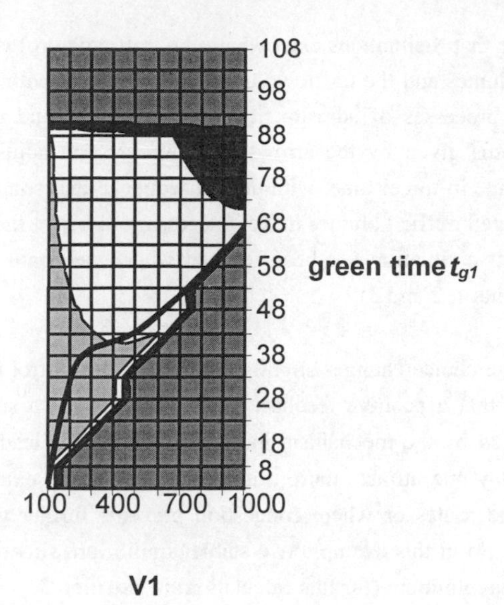

Figure 14 Iso-lines with equal values of the objective function *Z* and the (thick) line giving the equilibrium assignment for a given green time.

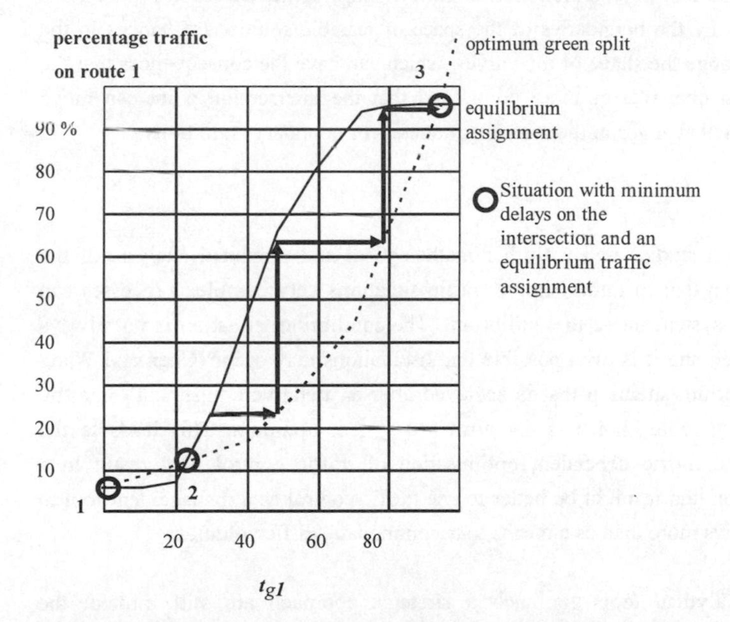

Figure 15 Optimum green time (dotted line) and equilibrium assignment (continuous line) with three equilibrium situations.

We see, in this example that 3 situations exist where the traffic control is optimised with respect to the traffic volumes and the traffic volumes are consistent with the travel times. If we assume that the processes of adjustment of traffic control and route choice are iterative, we find patterns given by the arrows in Figure 15: an adjustment in traffic control will give a change in travel time, with the consequence that some drivers choose another route. The changed traffic volumes make it necessary to adjust the control scheme etc. The process stops if a situation has been reached where the continuous and dotted curves intersect (i.e. points 1, 2 and 3).

If in situation 2 the route choice changes slightly and the traffic control is adapted to the changed flows, we see that a positive feedback mechanism exists: a small variation in route choice is reinforced by the mechanism in which more traffic leads to more green time which reduces delay and attracts more traffic etc. Only at the extremes, where all traffic chooses the same routes or where congestion prevents further growth, does the positive feed disappear. So in this example two stable equilibrium situations exist: 1 and 3. The total travel time is minimum (for this calculation) in situation 3.

Also in this case the form of the two curves of Figure 15 depends critically on control parameters and (saturation) flows. The optimal green split depends on minimum green time and the internal lost time, which means that the shape of the curves in Figure 15 is determined largely by the boundaries of the space of feasible solutions. Changes in the boundaries will change the shape of the curves, which can have the consequences that the curves intersect on one, two or more points and that the intersection point can move irregularly after small changes in the system parameters or boundary conditions.

CONCLUSION

In a few simulation studies and a further mathematical and graphical analysis of the problem it is shown that in rather simple traffic situations very complex processes can arise if we let the system move to equilibrium. The equilibrium situation is not always uniquely determined and it is even possible that oscillations may occur (Chen and Wang 1999). The equilibrium situation that is achieved after an iterative adjustment of traffic control to changing route choice is not always a system optimum. This leads to the conclusion that the traffic dependent optimisation of traffic control may result in a suboptimal situation and it might be better to use traffic control as a management tool to steer the traffic flows more than as a means to accommodate traffic volumes.

The necessary analytical tools for such a strategic approach are still limited: the combination of traffic control and traffic assignment that supports the search for a system optimum is still to be developed. Furthermore, the knowledge about the occurrence of the instabilities remains very limited at present. Empirical data on this subject may exist; there are even real time systems that estimate travel behaviour from real time traffic data (e.g. Bell and Grosso 1998). However, as far as the authors know, no analysis of the existing traffic data has been reported which looks for the existence of multiple stable equilibria or to the possibilities of increasing system performance by changing traffic control and route choice simultaneously.

The problem becomes more complex if we realise that in the real world the degrees of freedom for travellers are much larger than just route choice. The influence of traffic control on time of departure, modal choice, frequency of travelling, choice of destination etc. has been quantified by several researchers (Mokhtarian and Raney 1997, Mogridge 1997, Pendyala, Kitamura and Pas 1997)) but more should be done to apply these results to a method which optimises traffic control, taking into account the expected behavioural response. Apart from practical tools which make it possible to optimise traffic control, predict the impact on travel behaviour and to anticipate future changes in behaviour, there is also a need for an analytical framework to study the existence of equilibrium conditions in a system of traffic control and individual travellers.

REFERENCES

Allsop, R.E., 1974. Some possibilities for using traffic control to influence trip distribution and route choice, *Proceedings of the Sixth International Symposium on Transport and Traffic Theory*. New York: Elsevier.

Bell, M. G. H., S. Grosso, 1998. The Path Flow Estimator as a network observer. Traff. Engng. & Control Oct. 1998, pp 540 - 549.

Cascetta, E., M. Gallo and B. Montella, 1998. An asymmetric SUE model for the combined assignment-control problem. WCTR congers stream C3, Antwerp.

Charlesworth, J.A., 1977. The calculation of mutually consistent signal settings and traffic assignments for a signal controlled network. *Proceedings of the Seventh International Symposium on Transport and Traffic Theory*. Kyoto: The Institute of Systems Science Research pp. 545 – 569.

Chen H.-K. and C.-Y. Wang 1999. Dynamic Capacitated User-Optimal Route Choice Problem. TRB Conference 1999, Washington DC.

Fisk. C.S., 1984. Game theory and transportation system modelling. *Transpn. Res.* 18B pp 310 - 313

Mogridge, M.J.H., 1997. The self defeating nature of urban road capacity policy; a review of theories, disputes and available evidence, *Transport Policy* Vol. 4 Number 1 pp. 5 – 24.

Mokhtarian, P.L. and E.A. Raney, 1997. Behavioral response to congestion: identifying patterns and socio-economic differences in adaption. *Transport Policy* Vol. 4 Number 3 pp. 147 – 160.

Nakayama, S., R. Kitamura and S. Fujii, 1999. Drivers' Learning and Network Behaviour: A dynamic Analysis of Driver-Network System as a Complex System. TRB Conference presentation 990808, Washington DC.

H Pendyala, R.M.R. Kitamura, E.I. Pas, 1997. An Activity Based microsimulation analysis of transportation control measures. *Transport Policy* Vol. 4 Nummer 3 pp. 183 – 192.

H. Taale and F. Middelham, 1995. Simulating Traffic and Traffic Control with FLEXSYT-II,. Modelling and Simulation (M. Snorek, et al., Ed.). Society for Computer Simulation International, Istanbul, 1995, pp. 335-339.

H. Taale and F. Middelham, 1997. FLEXSYT-II-, A Validated Microscopic Simulation Tool,. Transportation Systems, Preprints of the 8th IFAC/IFIP/IFORS Symposium, Volume 2 (M. Papageorgiou and A. Pouliezos, Ed.). IFAC, pp. 923-928.

Taale and E. Scheerder. 1998. Paper for 5th World Congress on Intelligent Transport Systems, Seoul.

CHAPTER 6

PROBABILISTIC ROUTING AND SCHEDULING OF URBAN PICKUP/DELIVERY TRUCKS WITH VARIABLE TRAVEL TIMES

Eiichi Taniguchi
Department of Civil Engineering, Kyoto University
Yoshidahonmachi, Sakyo-ku, Kyoto 606-8501 Japan
taniguchi@urbanfac.kuciv.kyoto-u.ac.jp

Tadashi Yamada
Department of Civil Engineering, Kansai University
3-3-35 Yamate-cho, Suita, Osaka 564-8680 Japan
tyamada@ipcku.kansai-u.ac.jp

Dai Tamagawa
Hanshin Expressway Public Corporation
4-1-3 Kyutarocho, Chuo-ku, Osaka 541-0056 Japan

1 INTRODUCTION

This chapter presents a probabilistic vehicle routing and scheduling model that incorporates the variation of travel times. Dynamic traffic simulation was used to update the distribution of travel times. The model was applied to a test road network. Results indicated that the total cost decreased by implementing the probabilistic vehicle routing and scheduling model with variable travel times compared with the implementation of the forecast model.

2 CITY LOGISTICS INITIATIVES

There are many difficult problems concerning urban freight transport. Shippers and freight carriers are required to provide higher levels of service with lower costs to meet the various needs of customers. They have made efforts to rationalise their freight transport systems, but this has often led to an increase in pickup/delivery truck traffic in urban areas. The increase in the number of freight vehicles using urban roads has become a major source of traffic congestion, with many associated negative environmental impacts such as air pollution and noise. In addition, current global environmental agreements urge freight carriers to reduce CO_2 emissions produced from freight vehicles as well as passenger cars.

Some researchers (e.g. Ruske, 1994; Kohler, 1997; Taniguchi and van der Heijden, 2000) have proposed the idea of "City Logistics" to solve these difficult problems. The definition of City Logistics can be stated as, "the process for totally optimising the logistics and transport activities by private companies in urban areas considering the traffic environment, the traffic congestion and the energy savings within the framework of a free market economy". Although some of City Logistics initiatives listed below have only been proposed, others have already been implemented in several cities.

(a) Advanced information systems (e.g. Kohler, 1997; Taniguchi *et al.*, 1998b)
(b) Co-operative freight transport systems (e.g. Ruske, 1994; Taniguchi *et al.*, 1995)
(c) Public logistics terminals (e.g. Janssen and Oldenberger, 1991; Duin, 1997; Taniguchi *et al.*, 1999)
(d) Load factor controls (Taniguchi and van der Heijden, 2000)
(e) Underground freight transport systems (e.g. Koshi *et al.*, 1992; Visser, 1997; Duin, 1998; Ooishi and Taniguchi, 1999).

This chapter focuses on advanced information systems among these five initiatives. Advanced information systems have become important in rationalising existing logistics operations. In general, advanced information systems for pickup/delivery trucks operations have three important functions:

(a) To allow drivers and the control centre to communicate with each other
(b) To provide the real time information on the traffic conditions
(c) To store detailed historical pickup/delivery trucks operations data

The third function has not been fully discussed in the literature, but it is very important for rationalising logistics operations. A Japanese milk producing company experienced one successful application of historical operations data. After introducing a satellite based information system for one year, the company was able to reduce the number of pickup/delivery trucks by 13.5% (from 37 to 32 vehicles) and increase their average load factor by 10 percent (from 60% to 70%). A computer-based system was used to store detailed historical data of the pickup/delivery truck operations, including starting/arriving times at the depot and customers as well as the waiting times, travelling speeds and routes travelled. The company was able to analyse this data and change their routes and schedules to substantially increase the efficiency of their vehicle fleet. This type of system can reduce both freight transport and environmental costs within a city.

This chapter explores vehicle routing and scheduling procedures using advanced information systems in urban areas. Freight carriers have depots and their pickup/delivery trucks depart from the depot and visit customers within a designated time window for collecting or delivering goods and then return to the depot. This chapter presents a probabilistic vehicle routing and scheduling model with dynamic traffic simulation, which explicitly incorporates variable travel times. The uncertainty of travel times affects the identification of the optimal routes and schedules of pickup/delivery trucks on very congested urban roads. Recently the implementation of advanced information systems allows freight carriers to use historical and real time, travel time data on urban roads. A model is developed to quantify the benefits of considering the uncertainty of travel times in order to rationalise logistics systems and reduce the negative impacts of goods movement on the environment.

A number of operations researchers (e.g. Solomon, 1987; Koskosidis *et al.*, 1992; Russell, 1995; Bramel *et al.*, 1996; Taniguchi *et al.*, 1998a) have investigated vehicle routing problems with time windows (VRP-TW). Other researchers have studied stochastic vehicle routing and scheduling problems (e.g. Jaillet and Odoni, 1988; Dror *et al.*, 1989; Powell *et al.*, 1995; Gendreau *et al.*, 1996). Most research in this area has focused on dynamic routing and scheduling that considers the variation in customer demands. However, there has been limited research on routing and scheduling with probabilistic travel times.

3 The Probabilistic and Forecast Model

3.1 Framework

Figure 1 presents a framework of the model presented in this study. The model is composed of two sub-models; (a) a model for the probabilistic vehicle (pickup/delivery truck) routing and scheduling problem with time windows (P-VRP-TW) for each company and (b) a dynamic traffic simulation model for the fleet of pickup/delivery trucks and passenger cars on the road network within the city.

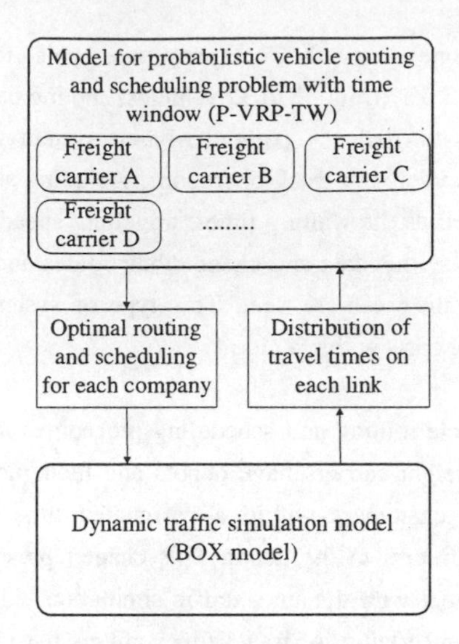

Figure 1 Framework of the model.

The model for P-VRP-TW is defined as follows. A depot and a number of customers are defined for each freight carrier. A fleet of identical vehicles collects goods from customers and delivers them to the depot or delivers goods to customers from the depot. For each customer a time window, specifying the desired time period to be visited, is also designated. For example, in the case of collecting goods, vehicles depart from the depot and visit a subset of customers for picking up goods in sequence and return to the depot to unload them. A vehicle is allowed to make multiple trips per day. Each customer must be assigned to exactly one route of a vehicle and all the goods from each customer must be loaded on the vehicle at the same time. The total weight of the goods in a route must not exceed the capacity of the vehicle. This problem is used to determine the optimal assignment of vehicles to customers and the departure time as well as the order of visiting customers for a freight carrier. P-VRP-TW explicitly incorporates the distribution of travel times for identifying the optimal routes and departure times of vehicles.

The optimal assignment of vehicles to customers and the departure time as well as the visiting order of customers for each freight carrier becomes input to the dynamic traffic simulation model. The dynamic traffic simulation model is based on a macroscopic dynamic simulation BOX model (Fujii *et al.*, 1994). This model estimates the distribution of travel times on each link in 1-hour intervals. The P-VRP-TW model is then re-solved using the updated distribution of travel times on each link obtained from the BOX model. Thus, the distribution of travel times for each link is represented by a normal distribution, in 1-hour time intervals. The model therefore, incorporates time dependent travel times.

3.2 P-VRP-TW Model

This section describes a mathematical model of the P-VRP-TW introduced in the previous section. The model minimises the total cost of distributing goods with truck capacity and designated time constraints. The total cost is composed of three components: (a) fixed cost of vehicles, (b) vehicle operating cost, which is proportional to time travelled and spent waiting at customer's place, (c) delay penalty for designated pickup/delivery time at customer's place.

Let,

$C(t_0, \mathbf{X})$: total cost (yen)

t_0: departure time vector for all vehicles at the depot

$$t_0 = \{t_{l,0} \mid l = 1, m\}$$

$\mathbf{X}$: assignment and order of visiting customers for all vehicles

$$\mathbf{X} = \{\mathbf{x}_l \mid l = 1, m\}$$

$\mathbf{x}_l$: assignment and order of visiting customers for vehicle l

$$\mathbf{x}_l = \{n(i) \mid i = 1, N_l\}$$

$n(i)$: node number of i th customer visited by a vehicle

N_l: total number of customers visited by vehicle l

m : maximum number of vehicles available

$c_{f,l}$: fixed cost for vehicle l (yen /vehicle)

$\delta_l(\mathbf{x}_l)$: = 1; if vehicle l is used

= 0; otherwise

$C_{t,l}(t_{l,0}, \mathbf{x}_l)$: operating cost for vehicle l (yen)

$C_{p,l}(t_{l,0}, \mathbf{x}_l)$: penalty cost for vehicle l (yen)

$c_{t,l}$: operating cost per minute for vehicle l (yen /min)

$t_{l,n(i)}$: departure time of vehicle l at customer's node $n(i)$

$\bar{T}(\bar{t}_{l,n(i)}, n(i), n(i+1))$: average travel time of vehicle l between customer's node $n(i)$ and $n(i+1)$ at time $\bar{t}_{l,n(i)}$

$t_{c,n(i)}$: loading/unloading time at customer's node $n(i)$

$p_{l,n(i)}(t_{l,0},t,\mathbf{x}_l)$: probability in which a vehicle that departs the depot at time $t_{l,0}$ arrives at customer's node $n(i)$ at time t

$c_{d,n(i)}(t)$: delay penalty cost per minute at customer's node $n(i)$ (yen/min)

$c_{e,n(i)}(t)$: early arrival penalty cost per minute at customer's node $n(i)$ (yen/min)

$W_l(\mathbf{x}_l)$: load of vehicle l (kg)

$W_{c,l}$: capacity of vehicle l (kg).

Then the model can be formulated as follows.
Minimise

$$C(t_0,\mathbf{X})=\sum_{l=1}^{m}c_{f,l}\cdot\delta_l(\mathbf{x}_l)+\sum_{l=1}^{m}E\left[C_{t,l}(t_{l,0},\mathbf{x}_l)\right]+\sum_{l=1}^{m}E\left[C_{p,l}(t_{l,0},\mathbf{x}_l)\right] \tag{1}$$

where,

$$E\left[C_{t,l}(t_{l,0},\mathbf{x}_l)\right]=c_{t,l}\sum_{i=0}^{N_l}\left\{\overline{T}(\bar{t}_{l,n(i)},n(i),n(i+1))+t_{c,n(i+1)}\right\} \tag{2}$$

$$E\left[C_{p,l}(t_{l,0},\mathbf{x}_l)\right]=\sum_{i=0}^{N_l}\int_0^{\infty}p_{l,n(i)}(t_{l,0},t,\mathbf{x}_l)\left\{c_{d,n(i)}(t)+c_{e,n(i)}(t)\right\}dt \tag{3}$$

Subject to

$$W_l(\mathbf{x}_l)\leq W_{c,l} \tag{4}$$

The problem specified by equations (1) - (4) is to determine the variable $\mathbf{X}$, that is, the assignment of vehicles and the visiting order of customers and the variable t_0, the departure time of vehicles from the depot. Note that $n(0)$ and $n(N_l+1)$ represent the depot in equations (2) and (3).

The distribution of travel times is required in equation (1) for determining the expected value of operating costs and penalty costs. This is the major difference of the probabilistic VRP-TW model from the forecast VRP-TW model. The forecast VRP-TW model adopts one number for the forecast travel time instead of the whole distribution. The dynamic traffic simulation calculates the distribution of travel times, which can be approximated by the normal distribution for every hour. Then the updated normal distribution can be input to the probabilistic VRP-TW model.

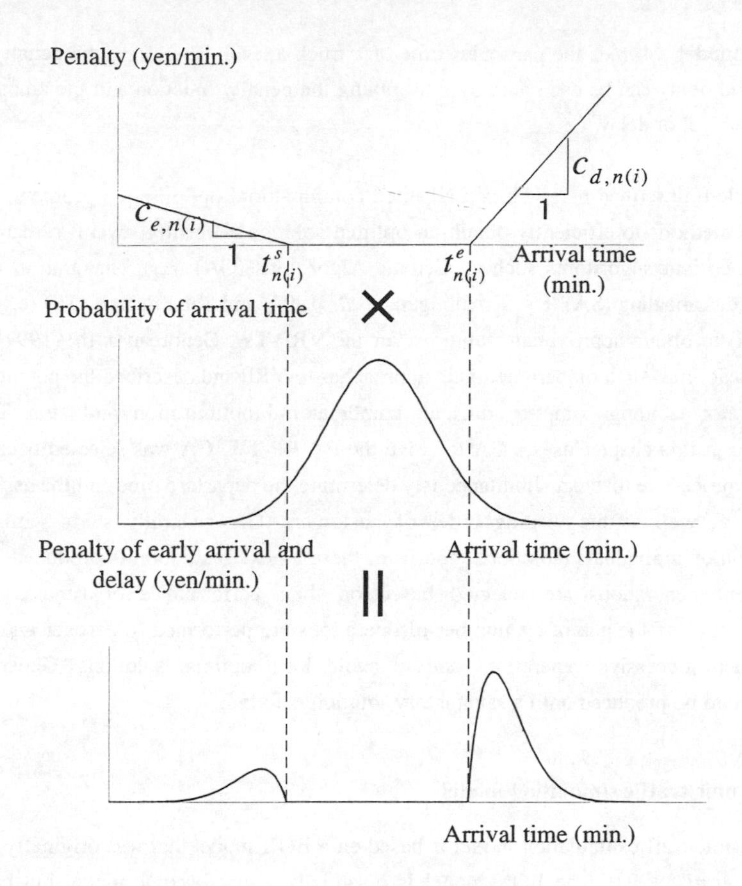

Figure 2 Penalty of early arrival and delay at customer's place for the probabilistic model.

Figure 2 shows the penalty for vehicle delay and early arrivals at customer's place. The figure demonstrates that the penalty function multiplied by the probability of arrival times is the penalty of early arrival and delay. The time period $(t^e_{n(i)} - t^s_{n(i)})$ of the penalty function defines the width of the soft time window in which vehicles are requested to arrive at customer's place. If a vehicle arrives at customer's place earlier than $t^s_{n(i)}$, it must wait until the start of the designated time window and a cost is incurred during waiting. If a vehicle is delayed, it must pay a penalty proportional to the amount of time it was delayed. This type of penalty is typically observed in goods distribution to shops and supermarkets in urban areas. Multiplying the penalty function and the probability of arrival time as shown in Figure 2 can identify the penalty of early arrivals and delay at customer's places for the probabilistic model. The forecast model assumes the particular time of a truck arrival. Therefore, the penalty for early arrival and delay can be estimated by multiplying the penalty function and the amount of time in early arrival or delay.

The problem described herewith is a NP-hard combinatorial optimisation problem. It requires heuristic methods to efficiently obtain an optimal solution. Recently several researchers have applied heuristic algorithms such as Genetic Algorithms (GA) (e.g. Thangiah *et al.*, 1991), Simulated Annealing (SA) (e.g. Kokubugata *et al.*, 1997) and Tabu Search (TS) (e.g. Potvin *et al.*, 1996) to obtain approximate solutions for the VRP-TW. Gendreau *et al.* (1997) reviewed the application of such modern heuristic approaches to VRP and described the potential of such methods for tackling complex, difficult combinatorial optimisation problems. The model described in this chapter uses a GA to solve the P-VRP-TW. GA was selected because it is a heuristic procedure that can simultaneously determine the departure time and the assignment of vehicles as well as the visiting order of customers. GA generally starts with an initial population of individuals (solutions) and from these a next generation is produced. Parents of subsequent generations are selected based on their performance or fitness. Using the characteristics of the parents, a number of operations are performed (crossover and mutation) to produce successive generations and to avoid local optimal solutions. Generations are continued to be produced until a satisfactory solution is found.

3.3 Dynamic traffic simulation model

The dynamic traffic simulation model is based on a BOX model that was originally developed by Fujii *et al.* (1994). The BOX model is essentially a macroscopic model but because the origin and destination of each vehicle are defined, it is actually a hybrid macroscopic-microscopic model. Vehicles are assumed to choose the shortest path when they arrive at a node using an estimated average travel time. The BOX model consists of two components, flow simulation and route choice simulation as shown in Figure 3. A sequence of boxes is used to represent each link. Groups of vehicles flowing out of a box and into the next box during the scanning interval represent the flow on links. This type of flow simulation is similar to the cell transmission model developed by Daganzo (1994). There are two assumptions for modelling links; (a) the maximum flow during a scanning interval is the same for all sections on links, (b) no inflow and outflow is allowed in the middle of links. A consequence of assumption (a) is that only the last section of a link can be a bottleneck, where a congestion queue starts. Two states of flow, congested flow and free flow, are represented.

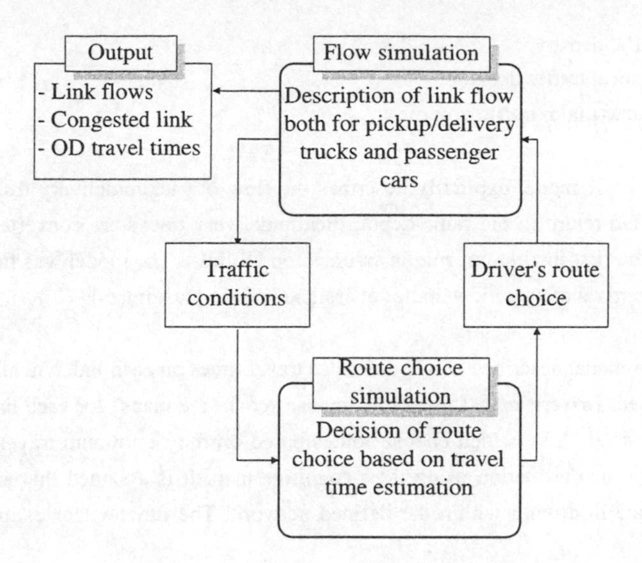

Figure 3 Structure of BOX model.

The time taken by a vehicle to proceed through a congested queue T_c is given by

$$T_c = \frac{F_c}{C_e} \tag{5}$$

where,

F_c : number of vehicles in a congestion queue

C_e : exiting traffic volume.

The exiting traffic volume is the traffic volume that can flow out of the last section of a link into the successive link. The time that is required to go through the running area without any queue T_f is estimated by

$$T_f = \frac{L_f}{V_f} \quad ; \text{if} \quad K \leq K_0 = \frac{Q_{\max}}{V_f} \tag{6}$$

$$T_f = \frac{L_f \cdot K}{Q_{\max}} \quad ; \text{if} \quad K > K_0 = \frac{Q_{\max}}{V_f} \tag{7}$$

where,

L_f : length of flowing area without any queue

V_f : free running speed

K : traffic density

K_0 : critical traffic density

Q_{max} : maximum traffic volume.

The modified BOX model explicitly describes the flow of pickup/delivery trucks that depart from a depot and return to the same depot. Pickup/delivery trucks are converted to passenger car units and the first-in-first-out rule is assumed on all links. The model was further modified to identify the arrival of specific vehicles at assigned nodes (customers).

The simulation model described above estimates travel times on each link and allows link costs to be determined. Drivers are assumed to compose "cognitive maps" for each link based on its estimated link cost. Drivers then choose routes based on their minimum travel cost from the current node to the destination using their cognitive map. It is assumed that all drivers have some experience in driving within the defined network. The function for estimating the link cost is:

$$C_k = T_{kt} + \eta_k \tag{8}$$

where,

C_k : estimated cost on link k

T_{kt} : travel time on link k at time t

η_k : disturbance term.

In this study the disturbance term η_k is assumed to be normally distributed with the zero mean and variance σ_η^2 as represented by

$$\eta_k \sim N(0, \sigma_\eta^2) \tag{9}$$

The travel times on each link vary within the day. The output of the BOX model is the updated distribution of travel times on each link. The distribution of travel times representing the interval of one hour was formulated by using the four-hour data on travel times. For example, the distribution of travel times representing the time interval 8:00-9:00 a.m. can be formulated by using the data between 6:30 – 10:30 a.m. Then this distribution was assumed to be normally distributed and the average and standard deviation were determined. The time interval of one hour was selected by checking the skewness and the kurtosis of the distribution.

4 BENEFITS OF USING PROBABILISTIC-VRP-TW MODEL

4.1 Test conditions

The model described in the previous section was applied to a test network with 25 nodes and 40 links as shown in Figure 4. This road network is comprised of the same type of roads with free running speed of 40 km/h. Any node within the network can generate and attract passenger car traffic. These nodes are referred to as centroids and are also candidate nodes to be visited by pickup/delivery trucks. Ten freight carriers are assumed to operate a maximum of 12 pickup/delivery trucks in this network. Each freight carrier has one depot whose location is shown in Table 1. Three different types of trucks, having a capacity of 2, 4 and 10 tons respectively, can be used. However, up to four trucks of each type can only be operated by each carrier. The passenger car equivalence rates, operating costs and fixed costs for each type of pickup/delivery truck are based on results from recent studies of truck operations in Japan. The number of customers for each carrier was generated randomly between 14 and 22, as shown in Table 1. The actual nodes to be visited for each carrier were also determined randomly from all nodes in the network. The freight demand at each customer was determined based on the distribution of freight demand in Kobe City, Japan.

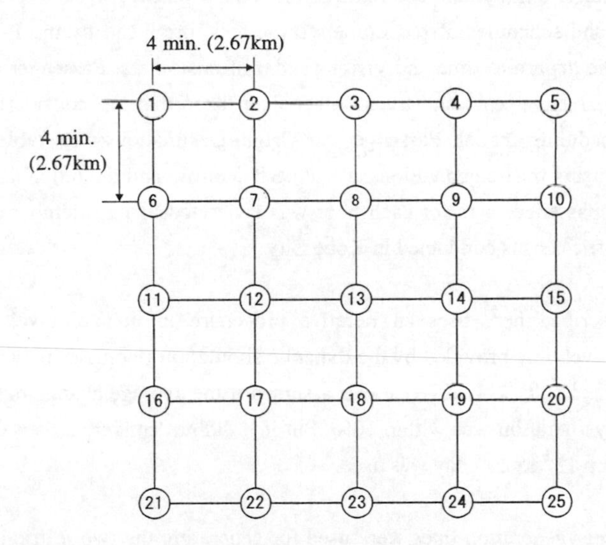

Figure 4 Test road network.

Table 1 Location of depot of 10 freight carriers.

Freight carrier	Depot node number	Number of customers
A	19	17
B	13	22
C	3	16
D	24	17
E	1	18
F	2	15
G	15	16
H	6	19
I	18	14
J	17	20

Three types of time windows were permitted in this study, time windows with one hour, time windows for a.m. (9:00-12:00) or p.m. (13:00-17:00), and no time window. The type and starting time of each customer's time window was based on the recent survey (1997) in the Kobe and Osaka area. The dynamic traffic simulation provides the distribution of travel times on each link for the scanning interval. In this study the scanning interval is 1 hour. When initially calculating the optimal routes and schedules, the average travel times on each link were assumed to be equal to the travel times using free running speeds.

The dynamic traffic simulation requires information on passenger car behaviour, as well as optimal routes and schedules of pickup/delivery trucks, produced by the P-VRP-TW model. This includes the departure time and visiting order of customers. Passenger cars in this study include actual passenger cars and trucks other than those that are considered in the optimal routing and scheduling model. Passenger car Origin-Destination (OD) tables for every hour were estimated using traffic generation rates at each centroid and the probability of O-D choice. The number of passenger cars for each hour was generated using a temporal demand pattern based on the traffic census conducted in Kobe City.

The model described here uses an iterative procedure to update travel time every day. Therefore the travel time provided by the dynamic simulation fluctuates between different days. In calculating the results, convergence was assumed if the average change of travel times from the previous days iteration was within 10%. But if it did not converge, then the average of the 10 iterations from 11th to 20th day was used.

Two passenger car generation rates were used for generating the two traffic conditions, that is (a) congested and (b) less congested. The total generation of passenger cars per day for case (a) was 260 (vehicle/day) and for case (b), 175 (vehicle/day) at each node. This generation was uniformly located at all nodes.

4.2 Results

Table 2 shows the comparison of the costs of freight carriers with the probabilistic VRP-TW model and the forecast VRP-TW model. The forecast VRP-TW model employs the average travel time for identifying the optimal routes and schedules of pickup/delivery trucks. This table shows that the total cost decreased by 1.44% by implementing the probabilistic VRP-TW in case (a), but it was slightly increased in case (b). Therefore, the probabilistic model can contribute to reduce the costs when the traffic is more congested and the travel time is more uncertain. The average travel speed was about 30 km/h in case (a) and 39 km/h in case (b), which was close to the free running speed of 40 km/h.

Table 2 Comparison of costs between the probabilistic and forecast VRP-TW.

case	model	fixed cost	operation cost	penalty for early arrival	delay penalty	total	change from base case
(a) congested	forecast (yen/day)	388,895	207,997	39,290	177,373	813,555	89,085
	probabilistic (yen/day)	404,376	213,742	38,709	145,021	801,848	77,378
	change (%)	3.98	2.76	-1.48	-18.24	-1.44	
(b) less congested	forecast (yen/day)	378,335	175,148	37,306	133,681	724,470	base case
	probabilistic (yen/day)	377,607	181,443	40,637	127,963	727,651	
	change (%)	-0.19	3.59	8.93	-4.28	0.44	

A comparison of the total costs in cases (a) and (b) is given in Table 2 by taking the forecast VRP-TW of case (b) as base case. The total cost of the forecast and probabilistic VRP-TW increased by 89,085 yen/day and 77,378 yen/day from base case, respectively. Therefore, the probabilistic VRP-TW resulted in a reduced increase of 11,707 yen/day from base case, which is 13.1% smaller than the increase of forecast VRP-TW. This reduction in the expected cost increase is a benefit of implementing the probabilistic VRP-TW when the traffic network became more congested. Table 3 compares the total cost for each freight carrier. In case (a), 7 freight carriers out of 10 enjoyed a reduction in total costs by using the probabilistic VRP-TW. In case (b), 5 freight carriers still achieved a reduction in total costs.

Table 2 also shows a considerable decrease in the delay penalty for the probabilistic model. The probabilistic model allows freight carriers to avoid delays at customer's place by taking into account the variable nature of travel times. In other words, it enables them to provide better service to customers by arriving within the collection/delivery time window. The fixed cost and operation cost in case (a) increased for the probabilistic VRP-TW, because more 10-ton trucks and less 2-ton and 4-ton trucks were used in the probabilistic case, as shown in Table 4. Choosing 10 ton-trucks by the freight carriers led to an increase in the number of customers visited by a vehicle on its single trip, departing and returning to the depot.

Table 5 shows the change of total travel time on the network. This table indicates that the total travel time of trucks was reduced by 1.15% in case (a) by adopting the probabilistic VRP-TW, but that of passenger cars slightly increased and, overall, the travel time of all vehicles increased by 0.33%. If we take the subtotal (truck) travel time of the forecast VRP-TW in case (b) as base case, then the increase in travel time of trucks was 840 min./day and 764 min./day for the forecast and probabilistic VRP-TW from base case, respectively. Therefore, the probabilistic VRP-TW had a lower increase (76 min./day), which is a 9.0% reduction compared with the forecast case, when the traffic network became congested.

Table 3 Comparison of total costs of each freight carrier.

(a) congested				(b) less congested			
freight carrier	forecast (yen/day)	probabilistic (yen/day)	change (%)	freight carrier	forecast (yen/day)	probabilistic (yen/day)	change (%)
A	89,963	87,293	-2.97	A	89,423	80,445	-10.04
B	105,683	96,951	-8.26	B	92,802	89,111	-3.98
C	59,870	63,089	5.38	C	60,312	60,121	-0.32
D	84,265	82,165	-2.49	D	74,977	70,440	-6.05
E	66,800	83,114	24.42	E	57,880	69,497	20.07
F	62,647	62,260	-0.62	F	57,982	58,735	1.30
G	79,562	77,883	-2.11	G	72,039	69,608	-3.37
H	114,963	109,132	-5.07	H	99,824	101,118	1.30
I	58,556	60,719	3.69	I	48,212	50,736	5.24
J	91,246	79,243	-13.15	J	71,019	77,840	9.60
total	813,555	801,849	-1.44	total	724,470	727,651	0.44

Table 4 Average number of trucks used in operation.

case	capacity of truck (ton)	forecast	probabilistic	change (%)
(a) congested	2	6.9	6.6	-4.35
	4	9.8	9.4	-4.08
	10	14.8	16.4	10.81
	total	31.5	32.4	2.86
(b) less congested	2	6.9	5.2	-24.64
	4	10.2	9.4	-7.84
	10	13.7	15.6	13.87
	total	30.8	30.2	-1.95

Table 5 Total travel time on the network.

case	model	passenger car	truck		subtotal (truck)	change from base case	total
			running time	waiting time			
(a) congested	forecast (min./day)	69,505	4,592	2,037	6,629	840	76,134
	probabilistic (min./day)	69,832	4,535	2,018	6,553	764	76,385
	change(%)	0.47	-1.24	-0.93	-1.15		0.33
(b) less congested	forecast (min./day)	35,989	3,690	2,099	5,789	base case	41,778
	probabilistic (min./day)	36,067	3,708	2,126	5,834		41,901
	change(%)	0.22	0.49	1.29	0.78		0.29

5 Conclusions

This chapter presented a probabilistic vehicle routing and scheduling model with variable travel times obtained from dynamic traffic simulation. This model explicitly incorporates the variation of travel times for identifying the optimal routes and schedules of pickup/delivery trucks in urban areas. The model was applied to a test road network and the following findings were derived.

(a) Implementing the probabilistic VRP-TW model decreased the increase in total costs due to congestion by 13.1% compared with the forecast VRP-TW model.

(b) Incorporating the variation of travel times considerably reduces the delay penalty. This enables freight carriers to provide better service to customers by improving the reliability to arrive at customer's place within the specified time windows.

(c) The increase in total travel time of pickup/delivery trucks due to congestion was also lower by implementing the probabilistic VRP-TW model by 9.0 % compared with the forecast VRP-TW model. This may be beneficial both for freight carriers and society at large.

REFERENCES

Bramel, J. and Simchi-Levi, D. (1996) Probabilistic analysis and practical algorithms for the vehicle routing problem with time windows. *Operations Research* **44**, 501-509.

Daganzo, C. F. (1994) The cell transmission model: A dynamic representation of highway traffic consistent with the hydrodynamic theory. *Transportation Research* **28B**, 269-287.

Dror, M., Laporte, G. and Trudeau, P. (1989) Vehicle routing with stochastic demands: properties and solution frameworks. *Transportation Science* **23**, 166-176.

Duin, J.H.R. van (1997) Evaluation and evolution of the city distribution concept. *Urban transport and the Environment for the 21st Century III*, WIT Press, pp. 327-337.

Duin, J.H.R. van (1998) Simulation of underground freight transport systems. *Urban transport and the Environment for the 21st Century IV*, WIT Press, pp. 149-158.

Fujii, S., Iida, Y. and Uchida, T. (1994) Dynamic simulation to evaluate vehicle navigation. *Vehicle Navigation & Information Systems Conference Proceedings*, pp.239-244.

Gendreau, M., Laporte, G. and Seguin, R. (1996) Stochastic vehicle routing. *European J. Oper. Res.* **88,** 3-12.

Gendreau, M., Laporte, G. & Potvin, J.-Y. (1997) Vehicle routing: modern heuristics, In E. Aarts & J. K. Lenstra (eds.), Chapter 9, *Local search in combinatorial Optimization,* John Wiley & Sons, pp.311-336.

Jaillet, P. and Odoni, A.R. (1988) The probabilistic vehicle routing problem. In B.L. Golden, A.A. Assad (eds.), *Vehicle routing; Methods and studies*, North-Holland, Amsterdam, pp. 293-318.

Janssen, B.J.P. and Oldenburger, A.H. (1991) Product channel logistics and city distribution centres; the case of the Netherlands. *OECD Seminar on Future Road Transport Systems and Infrastructures in Urban Areas,* pp.289-302.

Kokubugata, H., Itoyama, H. & Kawashima, H. (1997) Vehicle routing methods for city logistics operations, *IFAC/IFIP/IFORS Symposium on Transportation Systems*, Chania, Greece, eds. M. Papageorgiou & A. Pouliezos, pp.755-760.

Koshi, M., Yamada, H. and Taniguchi, E. (1992) New urban freight transport systems. *Selected Proceedings of 6th World Conference on Transport Research*, pp.2117-2128.

Koskosidis, Y.A., Powell, W.B. and Solomon M.M. (1992) An optimization-based heuristic for vehicle routing and scheduling with soft time window constraints. *Transportation Science* **26**, 69-85.

Kohler, U. (1997) An innovating concept for city-logistics. *4th World Congress on Intelligent Transport Systems*, Berlin, Germany, CD-ROM.

Ooishi, R. and Taniguchi, E. (1999) Effects and profitability of constructing the new underground freight transport system. In E. Taniguchi and R.G.Thompson (eds.) *City Logistics I,* Institute of Systems Science Research, Kyoto, pp.303-316.

Potvin, J.-Y., Kervahut, T., Garcia, B.-L. & Rousseau, J.-M. (1996) The vehicle routing problem with time windows; Part I: tabu search, *INFORMS Journal on Computing* **8**, 158-164.

Powell, W.B., Jaillet, P. and Odoni, A. R. (1995) Stochastic and dynamic network and routing. In M.O. Ball, T.L. Magnanti, C.L. Monma, G.L. Nemhauser (eds.). *Network Routing,* North-Holland, Amsterdam, pp. 141-295.

Ruske, W. (1994) City logistics --- Solutions for urban commercial transport by cooperative operation management. *OECD Seminar on Advanced Road Transport Technologies*, Omiya, Japan.

Russell, R.A. (1995) Hybrid heuristics for the vehicle routing problem with time windows. *Transportation Science* **29**, 156-166.

Solomon, M. M. (1987) Algorithms for the vehicle routing and scheduling problems with time window constraints. *Operations Research* **35**, 254-265.

Taniguchi, E., Yamada, T. and Yanagisawa, T. (1995) Issues and views on co-operative freight transportation systems. *7th World Conference on Transport Research,* Sydney.

Taniguchi, E., Yamada, T., Tamaishi, M. and Noritake, M. (1998a) Effects of designated time on pickup/delivery truck routing and scheduling. *Urban Transport and the Environment for the 21st Century IV*, WIT Press, pp.127-136.

Taniguchi, E., Thompson, R.G., and Yamada, T. (1998b) Vehicle routing and scheduling using ITS. *Proceedings of 5th World Congress on Intelligent Transport Systems,* Seoul, CD-ROM.

Taniguchi, E., Noritake, M., Yamada, T. and Izumitani, T. (1999) Optimal size and location planning of public logistics terminals. *Transportation Research* **35E(3)**, 207-222.

Taniguchi, E. and van der Heijden, R.E.C.M. (2000) An evaluation methodology for city logistics. *Transport Reviews* **20(1),** 65-90.

Thangiah, S. R., Nygard, K. E. & Juell, P. L. (1991) GIDEON: a genetic algorithm system for vehicle routing with time windows, *Seventh IEEE International Conference on Artificial Intelligence Applications*, IEEE Computer Society Press, Los Alamitos, CA, pp.322-328.

Visser, J.G.S.N. (1997) Underground networks for freight transport: a dedicated infrastructure for intermodal short-distance freight transport. *Proceedings of 3rd TRAIL Year Congress on Transport, Infrastructure and Logistics,* Delft, The Netherlands.

CHAPTER 7

THE N+M PERSON GAME APPROACH TO NETWORK RELIABILITY

Chris Cassir and Michael Bell
Transport Operation Research Group
University of Newcastle-upon-Tyne

INTRODUCTION

Game theory can provide some useful insights into the behaviour of a system offering several possibilities of action to its utility-maximising users. A transport network can be considered as such a system, where by and large all network users try to minimise their travel times. It is proposed in this chapter to use concepts from game theory in order to assess the reliability of transport networks in the face of random failures while taking into account the responses of users to the occurrence of such failures. First, we demonstrate that the traditional deterministic user equilibrium (DUE) assignment can be seen as the solution to a non-cooperative n-person game. Then we show that by adding a network tester as a player responsible for failures in the network, we obtain a game whose macroscopic solution can be found by solving an equivalent bi-level optimisation problem. The solution to this game can thus provide some lower-bound measures of network performance reliability since it reflects the robustness, from a cautious user's point of view, of the network to cope with particularly adverse circumstances. Finally we extend this methodology by allowing for non-homogeneity in the users' responses and solve a new bi-level problem, where the lower level problem represents a form of stochastic user equilibrium (SUE) assignment. Preliminary results are presented for a small real network.

BACKGROUND

A transport networks consists of a set of nodes and a set of links which together represent the transport infrastructure. The properties of the links and/or nodes may, however, be subject to random fluctuations which affect the cost of a trip as experienced by a network user. An extreme example would be an earthquake, which may cause some links to fail completely (with perhaps bridges or tunnels collapsing) and others to be significantly degraded in terms of the level of service they can offer the network user. A less extreme example may be a road accident or other incident causing the partial or total obstruction of a link.

This chapter is concerned with the reliability of transport networks as perceived by the network user. This is a function of both the reliability of the infrastructure components, the topology of the network, and the behaviour of the network users. For example, the consequences of a link failure for a network user depend on the existence of alternative routes, the state of knowledge of network users about both the failure and the route alternatives, and the reactions of the other network users.

In an earlier paper, Bell (2000) proposed a mental game as an approach to the measurement of network reliability. A network user "plays through" all that might go wrong on his trip. This may be formulated as a 2 person, zero sum, non-cooperative game between the network user, who is seeking the best route, and a network tester, who endeavours to hinder the network user as much as possible by causing individual links to fail. At the mixed strategy Nash equilibrium (see Nash, 1956), the network user is unable to reduce the expected trip cost by changing path choice probabilities while the network tester is unable to increase expected trip cost by changing link failure probabilities. There are two useful interpretations. The equilibrium trip cost may be regarded as that which would be expected by a pessimistic network user, and therefore may be regarded as an assessment of reliability from the user perspective. The equilibrium trip cost may also be regarded as the worst (from the user perspective) that can be expected by someone who has thoroughly tested the network, looking for the weakest links.

In Bell (2000), the 2 person, zero sum, non-cooperative game is formulated as a linear program, with or without path enumeration, and solved either by the simplex algorithm or the Method of Successive Averages. Where paths are enumerated, the constraint matrix consists primarily of the link-path incidence matrix and flow conservation takes the form of the summation of path choice probabilities to give link choice probabilities. Where paths are not enumerated, the constraint matrix consists primarily of the node-link incidence matrix and flow conservation is expressed at the node level (the probability of entering a node equals the probability of leaving it, unless the node is a source or sink).

A pessimistic estimate of travel cost will depend partially on pessimistic assumptions about the reactions of other network users. Taking the Newcastle Metro as an example, the thought process for a metro user trying to get from Jesmond to Tynemouth might be as follows: "If there were a delay at South Gosforth, my best option may be to change at the Monument. However, as other network users may be doing the same thing, causing congestion on the escalators, my best option may be to take a chance with South Gosforth. On the other hand, other network users may be thinking the same thing, etc.". The infinite chain of hypotheses typical of a game situation is evident here. As Bell (2000) considered only one network user, the reactions of others were not considered and link costs were treated as constant (on the basis that the marginal user would have a negligible impact on link costs).

The contribution of this chapter is to introduce other network users into the game. In this case, congestion must be considered and link costs no longer treated as constant. Unfortunately, the resulting minimax problem to be solved to find the Nash equilibrium is no longer linear or zero sum. In this chapter, a bi-level programming problem is formulated and solved for a test network by the Method of Successive Averages. The upper level of the problem represents the network tester, who is trying to maximise expected trip cost, while the lower level represents the response of the users, who react imperfectly to congestion in the network. For the lower level problem we use the logit path choice model, which allows for some dispersion in response and provides an explicit solution to the response problem.

TRAFFIC ASSIGNMENT AS AN N PERSON GAME

A traffic assignment can be viewed as a non-cooperative, n person game, where each network user endeavours to reach his destination by the best route possible. This is an n player game in the game theory sense because the payoff (the negative of trip cost) to each player (each network user) depends on the actions of the other players. We can then look at the correspondence between a macroscopic equilibrium assignment in the traditional (continuous) sense and a Nash equilibrium.

Proposition: a deterministic user equilibrium (DUE) with n users is equivalent to a mixed strategy Nash equilibrium for a non-cooperative, n person game, when n is large.

Demonstration: let us assume a network with one OD pair, several alternative routes, and a demand of n homogeneous users. At a deterministic user equilibrium, all paths used are minimum cost paths and all paths with greater than minimum cost are unused. In mathematical terms, this can be formulated as follows, defining **h** as a vector of route flows and **g** as the corresponding vector of route costs:

$$h_j = 0 \Leftarrow g_j(\mathbf{h}) > g_k(\mathbf{h}) \quad \text{for all routes } k \neq j$$

$$h_j > 0 \Rightarrow g_j(\mathbf{h}) = \min_k g_k(\mathbf{h}) = g_{od}(\mathbf{h})$$

Assuming n is large and using the Weak Law of Large Numbers, we can write $h_j \cong p_j n$, where p_j is the probability that route j is chosen by any user. Therefore at equilibrium $\mathbf{h}^*$, the conditions above imply that $p_j=0$ whenever $g_j(\mathbf{h}^*)$ is more than the minimum OD cost g_{od}.

At the microscopic level, if we now consider any user a among the total travelling between the same OD pair, we can then imagine choosing a route as equivalent to playing a game with all other users, where the aim of the game for all users is to minimise their individual travel costs. Allowing mixed route choice strategies (a mixed strategy s_a for user a is a convex combination of pure route choice strategies π_{aj} where $s_a=\Sigma_j \pi_{aj} p_{aj}$, with p_{aj} the probability that user a chooses route j), we can then define user a travel cost as follows, with respect to the n-vector $\mathbf{s}$ of mixed route strategies played by all users:

$$c_a(\mathbf{s}) = \Sigma_j p_{aj}\, c_{aj}(\mathbf{s}_{-a}) \quad (1)$$

where $c_{aj}(\mathbf{s}_{-a})$ is the expected cost to user a when he chooses route j and other users follow a set of strategies represented by the n-1 vector of mixed strategies $\mathbf{s}_{-a}$.

Since we assume that all users are homogeneous, it means that the cost or pay-off is perceived identically by all users in all situations. Since all players have the same alternatives (same routes) and they all try to maximise their pay-off, it follows that players can be permuted in the game. Thus $p_{1j} = p_{2j} = \ldots = p_{nj} = p_j$. Consequently $c_{1j}(\mathbf{s}_{-1}) = c_{2j}(\mathbf{s}_{-2}) = \ldots = c_{nj}(\mathbf{s}_{-n}) \cong c_j(\mathbf{s})$, if we assume that n is large so that the contribution of one user is negligible in the expected pay-off.

Since $p_{1j} = p_{2j} = \ldots = p_{nj} = p_j$, we can uniquely represent the vector of mixed route strategies $\mathbf{s}$ by the vector of route choice probabilities $\mathbf{p}$. Thus the expected cost of route j, $c_j(\mathbf{s})$, when users follow mixed strategies $\mathbf{s}$ equals the expected cost when users choose their routes according to the unique vector of route choice probabilities $\mathbf{p}$. If we define the cost of a route as being the route travel time, then because n is large and all users have the same route choice probabilities (resulting in deterministic flows equal to their expected values in the limit to infinity), we can approximate the expected cost by the travel time at expected route flow vector $\mathbf{h} = (p_j n)$:

$$c_j(\mathbf{s}) = c_j(\mathbf{p}) \cong g_j(n,\mathbf{p}) = g_j(\mathbf{h})$$

In summary, we have shown that $p_j = p_{aj}$, $\forall$ users a and $\forall$ route j, and also $c_{aj}(\mathbf{s}_{-a}) \cong c_j(\mathbf{p}) \cong g_j(\mathbf{h})$. It is now straightforward to see that the necessary and sufficient conditions for a deterministic user equilibrium are equivalent to the necessary and sufficient conditions for a Nash mixed strategy equilibrium:

$$p_j = 0 \Leftarrow g_j(\mathbf{h}) > g_{od}(\mathbf{h}) \Leftrightarrow p_{aj} = 0 \Leftarrow c_{aj}(\mathbf{s}_{-a}) > \min_k c_{ak}(\mathbf{s}_{-a})\ \forall\ a$$

(see Nash, 1951, replacing cost by utility and changing the signs accordingly). This result is easily extended to multi-commodity networks, although the players then have to be distinguished by their origins and destinations (users cannot be permuted across different OD pairs since their pay-offs differ). However, since the number of users for each OD pair is assumed to be known, the arguments developed in the single OD pair network case can be applied for each OD pair separately in a multi-commodity network, allowing of course for interactions between users of different OD pairs.

RELIABILITY AND THE N+1 PERSON GAME

We have shown in the previous section that a deterministic user equilibrium for one OD pair can be seen as a macroscopic, continuous equivalent to a Nash equilibrium in a non-cooperative n-person game. Following Bell (2000) we use this game concept and introduce another player into this game, a network tester who strives to minimise the pay-offs of all network users by damaging the network on one and only one link. This new player therefore has to play a mixed strategy with as many pure strategies as the number of links in the network. Damage on a link can range from minor incidents leading to some temporary reduction of the capacity to total link closure. For simplicity, we will arbitrarily consider that a link damaged by network tester action has its capacity reduced by half. By solving this n+1 person game and finding an equilibrium solution, we thus expect to obtain some measure of reliability or robustness of the network, taking into account the response of all users in a hazardous environment. The n+1 game can be formulated as follows:

G_1: Solve simultaneously
For each network user player a, $a \in (1,\ldots,n)$
$\text{Min}_{p_a} c_a(\mathbf{s},\mathbf{q}) = \Sigma_j p_{aj} \Sigma_k q_k\, c_{ajk}(\mathbf{s}_{-a})$

For the network tester player $n+1$:
$\text{Max}_{\mathbf{q}}\, c_{n+1}(\mathbf{s},\mathbf{q}) = \Sigma_k q_k\, c_{n+1,k}(\mathbf{s})$

Here $\mathbf{q}$ denotes the vector of scenario (or link damage) probabilities, $c_{ajk}(\mathbf{s}_{-a})$ is the cost to user a when choosing route j, in scenario k, and when $\mathbf{s}_{-a}$ occurs. $c_{n+1,k}(\mathbf{s})$ is the utility to the network tester on choosing scenario k when the n-vector of user's route strategies $\mathbf{s}$ occurs, and since the aim of this player is to increase the travel costs to all other players, we can define $c_{n+1,k}(\mathbf{s})$ as being the sum of all network user costs in scenario k:

$$c_{n+1,k}(\mathbf{s}) = \Sigma_a\, c_{ak}(\mathbf{s}) = \Sigma_a\, \Sigma_j\, p_{aj}\, c_{ajk}(\mathbf{s}_{-a}) \qquad (2)$$

According to Nash (1951), there is at least one mixed strategy equilibrium solution to this game. In practice it is, however, very difficult to find an equilibrium, especially when n gets large. This is why we propose to obtain an equivalent macroscopic solution to this game by solving the following bi-level optimisation problem:

B_1: Solve simultaneously

$$U: \ \text{Max}_{\mathbf{q}} \ \Sigma_j \Sigma_k \, q_k g_{jk}(\mathbf{h}) h_j \qquad \text{subject to} \qquad \Sigma_k \, q_k = 1, \ \mathbf{q} \geq 0$$

$$L: \ \text{Min}_{\mathbf{h}} \ \Sigma_u \Sigma_k q_k \int_0^{v_u(\mathbf{h})} t_{uk}(x) dx \qquad \text{subject to} \qquad \Sigma_j \, h_j = n, \ \mathbf{h} \geq 0$$

where link flow vector $\mathbf{v} = \mathbf{Ah}$, $\mathbf{A}$ is the link-route incidence matrix, and $t_{uk}(v_u)$ denotes the flow-dependent travel time on link u in scenario k.

Proposition: solving B_1 gives a macroscopic solution to the game G_1.

Demonstration: we can see that the upper level problem of B_1 solves the game from the network tester point of view, since it maximises total cost, which is equivalent to the sum of all user's costs. The lower level problem is a standard DUE assignment problem, except that it considers expected costs in a network subject to fluctuating conditions (inflicted by the network tester). At a joint optimum, the following equilibrium conditions apply:

Upper level: $\forall$ scenario k

$$q_k = 0 \ \Leftarrow \ \Sigma_j g_{jk}(\mathbf{h}) h_j < \max_r \Sigma_j g_{jr}(\mathbf{h}) h_j$$
$$q_k > 0 \ \Rightarrow \ \Sigma_j g_{jk}(\mathbf{h}) h_j = \max_r \Sigma_j g_{jr}(\mathbf{h}) h_j \qquad (3)$$

Lower level: $\forall$ route j

$$h_j = 0 \ \Leftarrow \ \Sigma_k g_{jk}(\mathbf{h}) q_k > \min_r \Sigma_k g_{rk}(\mathbf{h}) q_k$$
$$h_j > 0 \ \Rightarrow \ \Sigma_k g_{jk}(\mathbf{h}) q_k = \min_r \Sigma_k g_{rk}(\mathbf{h}) q_k \qquad (4)$$

Since the network users are permutable, by using the same idea presented in the previous section we can write $g_{jk}(\mathbf{h}) = c_{jk}(\mathbf{s}) \cong c_{ajk}(\mathbf{s}_{-a}) \ \forall a, \forall j, \forall k$ (the flow dependent route cost in any scenario is equal to the cost to any user in that scenario) and also $p_j = p_{aj}$ (the probability that route j is taken by any user is equal to the probability that any user a chooses route j). Thus we have for any scenario k:

$$\Sigma_j g_{jk}(\mathbf{h}) h_j = \Sigma_j \, c_{jk}(\mathbf{s}) \, h_j \cong \Sigma_j \, c_{ajk}(\mathbf{s}_{-a}) \, p_j \, n$$
$$= \Sigma_j \, n \, p_{aj} \, c_{ajk}(\mathbf{s}_{-a}) = \Sigma_j \Sigma_a \, p_{aj} \, c_{ajk}(\mathbf{s}_{-a})$$
$$= \Sigma_a \, \Sigma_j p_{aj} \, c_{ajk}(\mathbf{s}_{-a})$$
$$= c_{n+1,k}(\mathbf{s})$$

So (3) can be rewritten as follows: $\forall k$

$$q_k = 0 \Leftarrow c_{n+1,k}(\mathbf{s}) < \max_r c_{n+1,r}(\mathbf{s})$$
$$q_k > 0 \Rightarrow c_{n+1,k}(\mathbf{s}) = \max_r c_{n+1,r}(\mathbf{s}) \qquad (5)$$

Similarly we have, for any route j, provided we define the cost of route j to any user as the expected value of the cost across all scenarios:

$$\Sigma_k\, g_{jk}(\mathbf{h})\, q_k = \Sigma_k\, c_{jk}(\mathbf{s})\, q_k \cong \Sigma_k\, c_{ajk}(\mathbf{s}_{-a})\, q_k = E[c_{aj}]_{s\text{-}a} = c_{aj}(\mathbf{s}_{-a})$$

So, since $h_j = p_j\, n = p_{aj}\, n$, (4) can be rewritten : $\forall a \in (1,\ldots,n)$, $\forall j$

$$p_{aj} = 0 \Leftarrow c_{aj}(\mathbf{s}_{-a}) > \min_r c_{ar}(\mathbf{s}_{-a})$$
$$p_{aj} > 0 \Rightarrow c_{aj}(\mathbf{s}_{-a}) = \min_r c_{ar}(\mathbf{s}_{-a}) \qquad (6)$$

Conditions (5) and (6) are necessary and sufficient conditions for a Nash equilibrium of a non-cooperative, mixed strategy n+1 person game (see Nash, 1951). Thus we have shown that by solving B_1, we obtain an approximate macroscopic solution to the n+1 person game G_1 defined above, with the goodness of the approximation improving as n gets larger.

NON-HOMOGENEOUS NETWORK USERS

So far we have assumed that all network users are homogeneous, meaning that they perceive travel costs identically. This allowed us to establish the equivalence between a DUE and an n-person game (and also between B_1 and G_1), because network users are then interchangeable. However, in reality it is obvious that users do not perceive route costs identically, either because they have different levels of information or knowledge of the actual costs (say travel times) or because their utility functions are not identical. In a game context, this means that the pay-off associated to any route j for any user a cannot be expected to be the same as the pay-off to another user b for the same route under the same conditions. In other words, we do not have permutability any more. Since we cannot know the exact pay-off to each user, in order to formulate a game, we would have to assume some random distribution of c_{aj} applicable to any random user a. We can thus imagine a new n+1 person game where users do not know with certainty the pay-off to all other users in any situation. Although a formal proof is yet missing, we here propose to solve another bi-level optimisation problem that could establish a macroscopic correspondence with this new n+1 person game.

B_2: Solve simultaneously

Upper level: $\text{Max}_{\mathbf{q}}\ \Sigma_j\Sigma_k\, q_k g_{jk}(\mathbf{h}) h_j$ subject to $\Sigma_k\, q_k = 1,\ \mathbf{q} \geq 0$

Lower level: $\text{Min}_{\mathbf{h}}\ \Sigma_j h_j(\ln h_j - 1) + \alpha\left(\Sigma_u \Sigma_k q_k \int_0^{v_u(\mathbf{h})} t_{uk}(x)dx \right)$ subject to $\Sigma_j\, h_j = n$

At a joint optimum the following equilibrium conditions apply:

Upper level: $\forall k$

$$q_k = 0 \Leftarrow \Sigma_j g_{jk}(\mathbf{h})h_j < \max_r \Sigma_j g_{jr}(\mathbf{h})h_j$$

$$q_k > 0 \Rightarrow \Sigma_j g_{jk}(\mathbf{h})h_j = \max_r \Sigma_j g_{jr}(\mathbf{h})h_j \qquad (7)$$

Lower level: $\forall j$

$$h_j = n\, p_j(\mathbf{h}) = n \exp(-\alpha\Sigma_k g_{jk}(\mathbf{h})q_k) \,/\, \Sigma_r \exp(-\alpha\Sigma_k g_{rk}(\mathbf{h})q_k) \qquad (8)$$

The conditions (7) are the same as in B_1 and are equivalent to the network tester being unable to increase his pay-off by unilaterally changing his strategy. Conditions (8) describe a logit stochastic user equilibrium (SUE) for the flow vector **h**, with expected route costs depending on the network tester strategy. Even though it is difficult to formulate a game that has some equivalent Nash equilibrium conditions, we can, however, loosely interpret conditions (8) as representing an equilibrium in a game with non-identical users whose pay-offs are not known with certainty by both users. The result is a set of equilibrium route choice probabilities which, by the Weak Law of Large Numbers, determines the average flows and the corresponding expected average costs, allowing for some random perturbations (link failures) in the network aimed at penalising the users as much as possible. It is thought that a solution to B_2 can also provide some useful information about network reliability with the benefit of more realistic assumptions concerning the behaviour of network users (namely heterogeneity).

RELIABILITY GAME IN A MULTI-COMMODITY NETWORK

We should be able to apply the same principles for reliability analysis in a network that has several origin-destination (OD) pairs. There is, however, a question concerning the nature of the hypothetical game to be played in this case. Should we imagine a game played by n non-cooperative players spread over all OD pairs against one network tester, or should we rather envisage a series of parallel games, each played at the OD level between the OD users and an OD-specific network tester, while allowing for interactions in the network between all users? As we are interested in the performance of the network from the point of view of a user who chooses his route having "played" through all that might possibly go wrong for him on his trip, it arguably makes more sense to imagine one network tester for each OD pair.

We can thus imagine a new n+m game, with n non-cooperative heterogeneous network users spread across m OD pairs. Note that this implies a situation where a link may have failed for one OD but not for another. As we are dealing with hypothetical situations, this is not an inconsistency. Following the same logic used in the previous section, we propose to find a macroscopic solution to this game by solving the following problem:

B_3: Solve simultaneously for each OD pair $od \in (1,\ldots,p)$:

Upper level: $\mathrm{Max}_{\mathbf{q}od}\ \Sigma_j\Sigma_k\ q_{kod}\ g_{jkod}(\mathbf{h})h_{j,od}$ subject to $\Sigma_k\ q_{kod}=1,\ \mathbf{q}_{od}\geq 0$

Lower level: $\mathrm{Min}_{\mathbf{h}od}\ \Sigma_j h_{j,od}(\ln h_{j,od}-1) + \alpha(\ \Sigma_u\Sigma_k q_{kod}\int_0^{vu(\mathbf{h})} t_{uk}(x)dx\ )$ subject to $\Sigma_j\ h_{j,od} = n_{od}$

With $\mathbf{h}=(\mathbf{h}_1, \mathbf{h}_2,\ldots, \mathbf{h}_p)$ and $\Sigma_{od}\ n_{od} = n$. B_3 is a series of parallel bi-level problems linked by a common vector $\mathbf{h}$ of all route flows. The conditions for a global equilibrium are similar to those for B_3 except that the vector of link failure probabilities $\mathbf{q}$ is broken down into OD-specific sub-vectors $\mathbf{q}_{od}$.

ALGORITHM FOR SOLVING B_3

In order to solve problem B_3, we propose to use the Method of Successive Averages (MSA) which has proven reasonably efficient for solving the congested SUE problem. We will use the MSA for both $\mathbf{q}$ and $\mathbf{h}$, and proceed one step at a time for both the upper and lower level problem. For the lower level problem, we use the SUE assignment PFE (Path Flow Estimator), which has the benefit of generating and storing a set of partial paths as the procedure proceeds (the column generation technique). Although formal proof of the convergence for this bi-level problem is missing, we are confident that, when it does converge, it reaches a mutually consistent point (an equilibrium).

Step 0: Initialise m = 1 and $\mathbf{q}_{od}$ = 1 / number of links for all OD pairs
Step 1: Calculate expected link costs
Step 2: Obtain auxiliary link flows from the PFE
Step 3: Update link flows using the MSA
Step 4: Calculate expected scenario costs
Step 5: Obtain auxiliary scenario probabilities
Step 6: Update scenario probabilities by the MSA
Step 7: If convergence criteria are satisfied stop, otherwise set m := m+1 and return to *Step 1.*

EXAMPLE

The algorithm of the preceding section was tried on a medium–sized network of Leicester consisting of 103 links (including micro-links at junctions), 9 origins and 9 destinations. A representation of the network is shown in Fig.1 below:

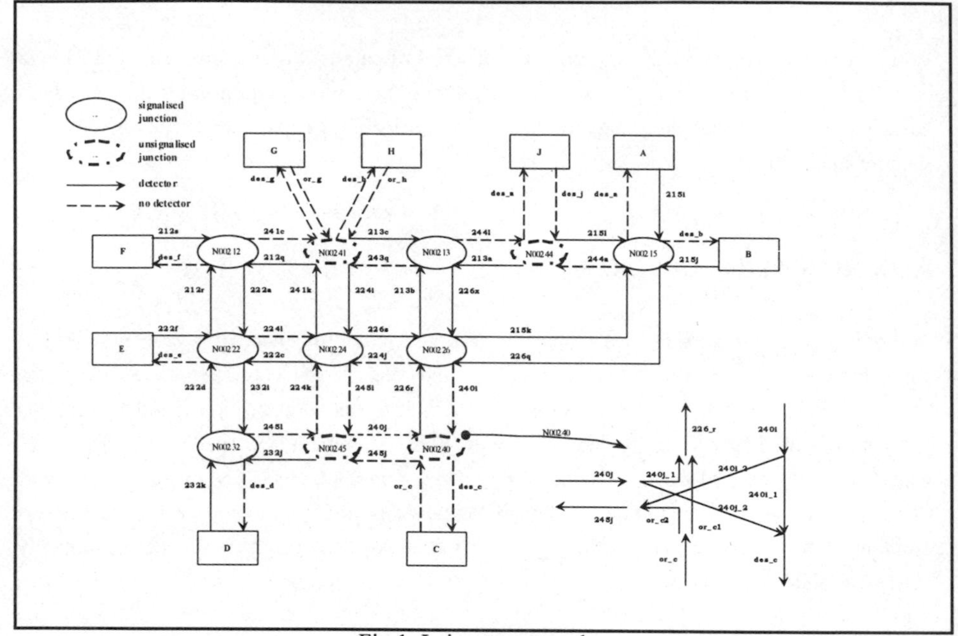

Fig. 1: Leicester network

Having obtained the OD-specific equilibrium link failure probabilities, it is then possible, knowing the equilibrium path flows, to calculate an OD reliability indicator as the probability that an OD user does not encounter a failed link on his trip when playing the hypothetical game described above. The table below gives those indicators for a sample of OD pairs, in ascending order of reliability, together with the corresponding link failure probabilities.

Table 1: Origin, destination, OD reliability and link failure probabilities

OR_C DES_D	R=0.3109	232J 0.72	222C 0.28		
OR_B DES_C	R=0.3603	226Q 0.79	226X 0.21		
OR_D DES_H	R=0.3611	226S 0.01	224K 0.72	224L 0.26	
OR_B DES_D	R=0.3650	226Q 0.78	222A 0.22		
OR_E DES_D	R=0.3942	232I 0.70	224L 0.30		
OR_E DES_C	R=0.4122	232I 0.67	224L 0.32		
OR_E DES_A	R=0.4334	232I 0.61	226R 0.01	226S 0.05	224L 0.32
OR_C DES_B	R=0.4465	215L 0.66	215K 0.33	224K 0.01	
OR_D DES_F	R=0.4684	212R 0.41	212Q 0.15	224K 0.44	
OR_E DES_F	R=0.4790	212R 0.61	212Q 0.01	224L 0.37	
OR_C DES_J	R=0.4939	226R 0.47	224K 0.52		
OR_C DES_H	R=0.5067	226R 0.53	224K 0.48		
OR_D DES_C	R=0.5236	226S 0.50	224K 0.01	224L 0.01	240J 0.48
OR_C DES_A	R=0.6432	215L 0.23	226R 0.44	226S 0.31	224K 0.03
OR_D DES_A	R=0.6553	215L 0.24	226R 0.34	226S 0.40	224K 0.03

This table is shown in order of increasing reliability to illustrate the type of results that can be obtained with the methodology presented above. We can see for instance that the method highlights those links, which for some specific OD pairs should be given special attention, since their failure will cause most damage to users of these OD pairs. It appears generally that the ODs which have a link with a relatively high probability of failure tend to be less reliable than ODs where link failure probabilities are more balanced, as one would expect since a high probability of failure indicates that the network tester has found a "weak link". Typically this would happen when a vulnerable (high failure probability) link attracts a big proportion of OD users even though the likelihood of failure is high.

CONCLUSIONS

This chapter demonstrates the correspondence between a deterministic user equilibrium traffic assignment and an n person game. This is extended to an n+1 person game by the introduction of a network tester. The correspondence between the Nash mixed strategy equilibrium of the n+1 person game and a macroscopic bi-level programming problem is shown. The assumed homogeneity of network users is then relaxed in two ways. Firstly, network users are no longer assumed to perceive costs identically and are therefore no longer treated as interchangeable. To allow for this, a stochastic user equilibrium route choice model is introduced into the bilevel formulation. Secondly, network users are no longer assumed to have the same origin and destination. Accordingly, extra network testers are added, leading to one for each OD pair and an n+m person game.

A method for the analysis of network reliability is presented and demonstrated for a real network. This method is conceptually derived from the notion of an n+m person, non-cooperative game being played between n imperfectly informed network users and m well informed network testers (one for each OD pair). The outcome of this game at equilibrium localises the vulnerable links for each OD pair and also determines a useful measure of OD reliability. Further work, on the theory side, will include the exact mathematical formulation of this game of incomplete information and its formal correspondence with the bi-level problem solved at the macroscopic level, and on the practical side, some further experience with reliability analysis based on this methodology.

REFERENCES

Bell, M.G.H (2000). A game theory approach to measuring the performance reliability of transport networks. *Transportation Research B*, Vol. 34B, 533-546.

Nash, J (1951). Non-cooperative games. *Annals of Mathematics.* Vol 54, pp 286-295.

CHAPTER 8

SNOWFALL WEATHER FORECAST AND EXPRESSWAY NETWORK RELIABILITY ASSESSMENT

Hiroshi Wakabayashi,
Faculty of Urban Science, Meijo University,
4-3-3,Nijigaoka, Kani, Gifu 509-0261, Japan.

1 INTRODUCTION

This chapter discusses expressway reliability assessment when the snowfall weather forecast is provided. First, the basic method using Reliability Graph Analysis (RGA) is presented. Second, consideration for dependent events is described. Two basic ideas are discussed on how to consider the dependent events. The first one involves introducing a correlation indicator into the RGA method. The second one is using travel time reliability. Next, data showing the relationship between the snowfall forecast and actual traffic control are discussed. Lastly, the actual reliability analysis without considering the dependent event is carried out.

2 DIFFICULTIES IN RELIABILITY ANALYSIS

This study aims to develop a reliability assessment method for bad weather, especially a snowfall environment. In developing a reliability assessment method, the first important point is its efficiency.

Many reliability assessment methods have been proposed so far. The computational work and memory size required, however, increases exponentially as the system expands. The computational work of any methods can be classified into four categories; that is, it is either proportional to 2^l, 2^p, 2^k or l^l, where l, p and k are the total number of links, paths, and cuts respectively, and I is an iteration number in the method proposed by Fratta and Montanary (1973). The highway network is a large system, thus an efficient method for the calculation of reliability should be developed.

The second important point is to consider inter-dependent events, when needed. Link failure caused by a traffic accident or a disaster is regarded as an independent event. However, a link failure caused by traffic congestion and bad weather is regarded as an inter-dependent event. Reliability assessment methods that incorporate inter-dependent events can assess the network reliability more accurately.

This chapter consists of seven sections including this one. Section three describes the significance of reliability and the motivation of this study. Section four explains the significance of reliability in more detail and presents the reliability assessment methods developed by the authors. The advantages of these methods are stated. The necessity for considering dependent events is also discussed. An approximate method using correlation indicator is then presented along with another method that considers travel time reliability. Section five shows the outline of the weather forecast in Japan, while section six describes the snowfall weather forecast and the expressway network reliability assessment briefly. Finally section seven presents the current conclusions of this study.

3 SIGNIFICANCE OF RELATIONSHIP BETWEEN NETWORK RELIABILITY AND SNOWFALL: MOTIVATION OF THE STUDY

The Japanese highway network suffered great damage from the 1995 Hyogo-ken Nanbu (Kobe) Earthquake. There is the potential for damage to the network's function, both from underground events, like an earthquake, and from the sky, like snowfall or heavy rain.

The core network is a trunk line between Tokyo (which has the largest population in Japan) and Osaka (second largest population) via Nagoya (third largest population). There is a region between Nagoya and Osaka that has a great potential for sudden snowfall in winter. There are two existing routes between Nagoya and Osaka. One route is a high standard arterial expressway. Although the other route is a highly standardized highway, it is just a fully access-controlled highway with lower limited speed than the expressway. The expressway with higher capacity was constructed through the snowfall area because the construction cost was cheaper than that for the route avoiding the snowfall area. However, when the expressway is closed due to snowfall, the traffic on the expressway is diverted to the surface highways, where there is already heavy traffic. Consequently, high traffic congestion occurs. Thus, it is very important to maintain the availability of the expressway. In addition, a new route avoiding the snow area is under construction.

The purpose of this study is to assess the reliability of this big network under a snowfall risk environment. In addition, when a new route is added to the existing network, the improvement in reliability should be evaluated. In this assessment, since many links are likely to be closed simultaneously, inter-dependent events should be considered.

4 METHODOLOGIES FOR NETWORK RELIABILITY ASSESSMENT

Significant of Reliability and Reliability Assessment Method

There are two types of reliability. The first is terminal (node to node) reliability. The second is travel time reliability (Wakabayashi and Iida, 1993).

The terminal reliability of the highway network is defined as the probability that two given nodes over the network are connected with a certain service level of traffic for a given time period, and it is obtained from the mathematical definition of Barlow and Proschan (1965). This reliability is an indicator of alternativity and redundancy.

When the traffic service level is smooth and without long delays, a road network with a high reliability level provides sure and unfluctuating traffic service by offering drivers alternative routes even when a certain route is closed. Thus, this reliability indicator illustrates the stability of the road network performance in terms of connectivity, in case of not only disaster but also traffic accidents or maintenance work. At the same time, this indicator can express the network conditions when diverted traffic, resulting from damage on some part of the network, affects the congestion in another part of the network. In other words, it represents the network service level that follows the failure in a network due to a disaster or an accident. This indicator, therefore, can be used in daily traffic management to reflect smooth transportation. In addition, a highly reliable network can provide the traffic service that ambulances and fire engines need in order to arrive at their destinations quickly and smoothly, and administer medical care, thus minimizing the impact of disasters.

Travel time reliability is defined as the probability that the travel time between two given nodes is guaranteed within a certain travel time. It can also be treated as the maximum travel time needed to arrive at the destination with a given probability. It is insufficient to provide drivers with the average travel time in advanced highway systems. Drivers are eager to know the degree of fluctuation of travel time, particularly when their trips are crucial. The indicator of travel time reliability is the rapidity and the accuracy of travel time. A road network that offers some measurements of travel time reliability can provide drivers with the estimated travel time to their destinations, for example, to an airport or a station within a given tolerance.

Methods for Reliability Assessment from Given Link Reliability

Boolean Method

First, the terminal reliability is defined. It is defined as the probability that two given nodes (*i, j*) in a network are connected, with a given service level of traffic and for a given time period. Similarly, link reliability is defined as the probability that the link can provide a certain service level for a given time period.

Terminal reliability, *R,* is given by the minimal path expression (Barlow and Proschan, 1965);

$$R = \mathrm{E}[1 - \prod_{s=1}^{p} (1 - \prod_{a \in Ps} Xa)], \tag{1}$$

and by the minimal cut expression;

$$R = \mathrm{E}[\prod_{s=1}^{k} \{1 - \prod_{a \in Ks} (1 - Xa)\}], \tag{2}$$

where P_s is the s-th minimal path set, and K_s is the s-th minimal cut set between node pair i and j, and p and k are the total number of minimal paths and cuts respectively. X_a is a binary indicator variable for link a, as follows:

$$Xa = \begin{cases} 1, \text{ if link } a \text{ provides the given traffic service level,} \\ 0, \text{ otherwise.} \end{cases} \tag{3}$$

Link reliability, *ra*, is

$$ra = \mathrm{E}[Xa]. \tag{4}$$

In Equation (1), the same variable, X_a, tends to appear more than once. To avoid using the product of the same indicator variable in the calculation of the expected value of reliability in Eq.(1), Boolean absorption, such as X_a*X_a=X_a (not X_a^2) is required. In addition, Eq. (1) needs all the minimal paths. These procedures require a great amount of computational work. The same argument also holds for Eq. (2).

When a minimal path is denoted as

$$\alpha_s = \prod_{a \in P_s} X_a \ , \tag{5}$$

Equation (1) can be rewritten by substitution;

$$R = \mathrm{E}[1 - (1 - \alpha_1)(1 - \alpha_2)\ldots(1 - \alpha_p)]\,. \tag{6}$$

Further expansion of Eq.(6) produces 2^p-1 terms where p is the number of minimal paths. Because Boolean absorption must be carried out for most of these terms, the calculation work is very time consuming. The CPU time and memory size required increases exponentially as the network size expands. For the actual network, computing reliability is very difficult or almost impossible.

If one path can be dropped, the calculation is approximately halved. Likewise, calculation can be reduced exponentially by decreasing the number of minimal paths. The minimal paths dropped are those paths whose contribution to the value of reliability is small, such as zigzag and detour paths. Those paths are not regarded as routes used by drivers and are meaningless for the calculation of reliability (Wakabayashi and Iida, 1992). When a partial paths set is employed in Eq.(1), it becomes

$$L_1 = \mathrm{E}[1 - \prod_{s=1}^{p'} (1 - \prod_{a \in P_s} X_a)]\,, \tag{7}$$

where L_1 is the lower bound of reliability proposed by Henley and Kumamoto (1981), and

$$p' \leq p\,, \tag{8}$$

where p' is the number of minimal paths in the partial set. If all the paths are used, that is, if $p' = p$, the value of L_1 obviously agrees with the exact value of reliability R in Eq.(1).

For minimal cuts, when a partial cuts is employed in Eq.(2), we have

$$U_1 = \mathrm{E}[\prod_{s=1}^{k'} \{1 - \prod_{a \in K_s} (1 - X_a)\}]\,, \tag{9}$$

where U_1 is the upper bound of reliability, and

$$k' \leq k\,, \tag{10}$$

where k' is the number of minimal cuts in the partial set. If all the cuts are used, that is, if $k' = k$, the value of U_1 obviously agrees with the exact value of reliability R in Eq.(2).

An efficient algorithm for calculating the value of reliability using this method has also been developed (Wakabayashi and Iida, 1992).

The Intersection Method

A more efficient approximation method using partial minimal path sets and cut sets is also proposed (Wakabayashi and Iida, 1993, 1994, Bell and Iida, 1997). Two reliability functions are introduced;

$$R_p = 1 - \prod_{s=1}^{p'} (1 - \prod_{a \in Ps} r_a)\,, \tag{11}$$

and

$$R_k = \prod_{s=1}^{k'} \{1 - \prod_{a \in Ks} (1 - r_a)\}\,, \tag{12}$$

where r_a is link reliability.

The value of Rp is characterized as follows: Rp is an increasing function of the number of paths. When the number of minimal path p' is small, Eq. (11) provides a lower bound value of reliability. When all the minimal paths are employed in Eq.(11), the value Rp yields an upper bound value of reliability. This upper bound is known as Esary and Proschan's upper bound, U_2, (Barlow and Proschan, 1965), and is

$$U_2 = 1 - \prod_{s=1}^{p} (1 - \prod_{a \in Ps} r_a)\,. \tag{13}$$

Hence, the value Rp increases monotonically with an increase in the number of paths from the lower bound of reliability to Esary and Proschan's upper bound of reliability. Similarly, the value of Rk decreases monotonically with an increase in the number of cuts from the upper bound of reliability to the lower bound of reliability. This lower bound of reliability, L_2, is known as Esary and Proschan's lower bound and is given by

$$L_2 = \prod_{s=1}^{k} \{1 - \prod_{a \in Ks} (1 - r_a)\}\,, \tag{14}$$

and

$$L_2 \leq R \leq U_2. \tag{15}$$

Therefore, the two functions, *Rp* and *Rk* will cross at a certain point between Esary and Proschan's upper and lower bounds. The value at the intersection has been proposed as an approximation for the reliability value.

Selection of the Partial Minimal Path Sets and Cut Sets: The *n*-th Shortest Route Search Problem

An approximate value of reliability depends greatly on how to select the partial minimal path set and cut set. This section describes an efficient way for selecting the partial minimal path set and cut set for a good approximation of the reliability value.

The reliability of a *s*-th minimal path *Ps* in Eq.(1) is

$$\Pr\{P_s\} = \prod_{a \in Ps} r_a. \tag{16}$$

Similarly, the unreliability of a *s*-th minimal cut *Ks* in Eq.(12) is

$$\Pr\{K_s\} = \prod_{a \in Ks} (1 - r_a). \tag{17}$$

Rp in Eq. (11) and *Rk* in Eq.(12) are increasing and decreasing functions of the number of paths and cuts respectively. Thus it is expected that we can get a good approximation of reliability for a small number of paths and cuts, if we successfully select the minimal path set and cut set which contribute most to the reliability value.

A logarithm of Eq.(16) leads to

$$\log(\prod r_a) = \log r_{a_1} + \log r_{a_2} + \ldots + \log r_{a_m}, \tag{18}$$

where m is the number of links included in this minimal path set. From the inequality,

$$0 \leq r_a \leq 1, \tag{19}$$

$-\log r_a$ can be regarded as a virtual link length. Thus the problem of finding the minimal path set with the highest reliability can be treated equivalently as the problem of finding the shortest route over the network. Once the shortest route is chosen, next, the second-shortest route should be chosen. Therefore the problem of selecting the partial minimal path set is reduced to the problem of finding the n-th shortest route over the network successively.

In the case of minimal cuts, a dual network representation is used since a minimal cut in the original network is exactly equivalent to the corresponding minimal path in the dual network. In the dual network, the link reliability is replaced by (1- ra) as the virtual link length for convenient computation work. For the selection of minimal cut set, the same procedure is followed for minimal path set over a dual network.

Advantage and Improvements for These Methods

These methods were firstly developed for assessing the quality of the network service, since the conventional evaluation methods were only used for assessing the network quantitatively. Reliability is one of the major and important qualitative network indicators.

The first advantage of these methods is practicality and convenience: conventional network reliability analysis requires in general a great amount of computational work and memory storage. When the network size expands, the CPU-time and memory size required increases exponentially. CPU time is proportionate to 2^l, where l is any indicator for the network size. If there is a computer with maximum capacity, and if there is a maximum size of highway network that can be handled, the calculation becomes impossible when only one link is added. The methods proposed by the authors are characterized as highly efficient methods. Secondly, the proposed methods are based on the reliability graph analysis (RGA) and can consider drivers' route choice principles. The original RGA method should consider all the minimal paths or cuts, which includes long detour routes.

The next step in this study is to consider inter-dependent events. Almost all conventional reliability assessment methods assume the independence of failure between the components of the system. The assumption of independence has permitted the further development of this study field. Many actual systems, however, show that some components have a dependent failure relationship. A reliability assessment method for a highway network system has the same property. Although assessment methods with inter-dependency have been studied, applicable cases are limited. In addition, they are not essential for considering intersystem or intercomponent dependencies. One example of such cases is where the components of the plant fail simultaneously due to an airplane crash. This case is not inter-dependent. We should consider the inter-dependent failure caused endogenously, not from the outside. At a job site, however, the dependent failure is considered by modifying the numerical result of the conventional method using the accumulated experiences and engineers' technological judgement. This suggests that a reliability analysis based on independence of the component's failures is useful in practice.

Requirements for Dependent Events Consideration

Inter-dependence of failures in a highway network is caused by traffic and exogenous factors. Highway availability under snowfall is such a case. There is a potential approximate approach. Development of the method should consider the following requirements.

1) As stated previously, the calculation for reliability is computationally time consuming. In some cases, it is impossible due to the size of network. Thus, it is important to guarantee the feasibility of the calculation.

2) The method should match the conventional reliability analyses that aim towards the feasibility of calculation.

3) The method should satisfy the boundary condition, or should include methodologies based on the assumption of intercomponent independence.

4) For practical use and calculation, the data source should be easy to access, or the method should be based on the data source availability.

Approximate Method using Correlation Indicator

Considering these requirements, we discuss the following reliability assessment method.

Let us consider a simple series system of two links. Let r_1=r_2=r, for the sake of brevity. When each link is independent in reliability, the reliability R_{AB} between node A and B is $R = r_1 \cdot r_2 = r^2$. If each link has maximum positive inter-dependence, $R = r$.

Based on the discussions of the previous two sub-sections, we consider

$$R_{AB} = r_1^{(1-\rho_{12}/2)} r_2^{(1-\rho_{12}/2)}, \tag{20}$$

where ρ is the correlation coefficient.

Following the same idea, for series system with n links, we would have

$$R_{AB} = r_1^{(1-f_1)} r_2^{(1-f_2)} \ldots r_n^{(1-f_n)}, \tag{21}$$

or

$$R_{AB} = \prod_{a,\text{for Independent Area}} r_a \prod_{a,\text{for Dependent Area}} r_a^{(1-f_a)}, \tag{22}$$

where f_a is the indicator for the correlation of link a with other links.

Similarly, in terms of minimal cuts, the reliability can be expressed as

$$R_{AB} = 1-(1-r_1)^{(1-g_1)}(1-r_2)^{(1-g_2)}...(1-r_n)^{(1-g_n)}, \tag{23}$$

or

$$R_{AB} = \{1- \prod_{a,\text{for Independent Area}} (1-r_a)\} \cdot \{1- \prod_{a,\text{for Dependent Area}} (1-r_a)^{(1-g_a)}\}. \tag{24}$$

Since it is difficult to observe the correlation indicators directly, the problem is how to estimate these indicators from other traffic characteristics in the network. In addition, this idea should be tested as an adequate approximate method.

This method is a simple method, but differs, however, from the conventional reliability analyses such as Boolean method, because the Boolean algebra contains logical sum and product in the calculation procedure.

This method was tested by using a traffic assignment simulation model with OD flow variation iteratively. The test network is a 9 by 9 rectangular type and the numbers of links and centroids are 288 and 81(=9^2), respectively.

A set of link reliabilities and the reliability between a pair of nodes can be obtained directly from the iterative simulation. At the same time, correlation of the link flows and link reliability between links can be obtained. Using Eqs.(25) and (26) with observed link reliability and correlation of link reliability or link flow, the reliability between a pair of nodes is calculated and compared with the observed reliability between the same node pair.

The following equations are used:

$$R_p = 1-\prod_{s=1}^{p'} (1- \prod_{a,\text{for Independent Area}} r_a \cdot \prod_{a,\text{for Dependent Area}} r_a^{(1-f_a)}), \tag{25}$$

and

$$R_k = \prod_{s=1}^{k'} \{1 - \prod_{a,\text{for Independent Area}} (1-r_a)\} \cdot \{1 - \prod_{a,\text{for Dependent Area}} (1-r_a)^{(1-g_a)}\}. \tag{26}$$

Method-1 (with minimal path sets, Eq. (25)), method-2 (with minimal cut sets, Eq. (26), and method-3 (the average value of method-1 and method-2) were calculated. Although method-2 gives a good approximation, this method is only an approximate method. Further study is needed.

Table 1. Results of the dependent events consideration of the test network.

OD Pair	Node to Node Reliability	Case1	Case2	Case3	Case4	Case5	Case6
	Direct Observation	1.0000	0.8400	0.7600	0.6400	0.2200	0.3800
#1 to #81	Method-1	1.0000	0.9720	0.9560	0.9040	0.7050	0.7470
	Method-2	(1.0000)	0.8370	0.7210	0.5920	0.1520	0.3920
	Method-3	1.0000	0.9045	0.8385	0.7480	0.4285	0.5695
	Direct Observation	1.0000	0.8200	0.7600	0.6200	0.2200	0.3600
#37 to #45	Method-1	1.0000	0.9450	0.9090	0.8060	0.4340	0.5630
	Method-2	(1.0000)	0.8380	0.7210	0.6180	0.1550	0.3920
	Method-3	1.0000	0.8915	0.8150	0.7210	0.2945	0.4775
	Direct Observation	1.0000	0.8400	0.7800	0.6600	0.2400	0.4000
#5 to #77	Method-1	1.0000	0.9610	0.9810	0.8440	0.6930	0.6530
	Method-2	(1.0000)	0.8230	0.6760	0.5710	0.1160	0.4070
	Method-3	1.0000	0.8920	0.8285	0.7075	0.4045	0.5300

Travel Time Reliability

Another approach for inter-dependent failure is to use travel time reliability. As stated in Section 3.1, travel time reliability is defined as the probability that the travel time between two nodes is less than a given travel time. It can be also treated as the maximum travel time to arrive at the destination with a given probability.

First, variation of traffic volume is assumed to be dominant to travel time variation. From the condition that the probability in terms of traffic volume is the same as the probability in terms of travel time for the same traffic condition, we have

$$h_a(t) = f_a(v)dv/dt, \tag{27}$$

where $h_a(t)$ is a probability density function for travel time on link a; and
$f_a(v)$ is a probability density function for demand traffic volume.

Using B.P.R. function as the link performance function,

$$t_a(v) = t_{a0}[1 + \alpha(v_a / C_a)^{\beta}], \tag{28}$$

where $t_a$and t_{a0} are the travel time and the free-flow travel time respectively on link a, and the quantities α and β are model parameters. By substitution,

$$h_a(t) = f_a(v)C_a^{\ \beta} / (\alpha\beta t_{a0} v_a^{\ \beta-1}). \tag{29}$$

From Eq. (29), probability distribution function $H_a(t)$ is obtained as

$$H_a(t) = \int_0^t h_a(t)dt. \tag{30}$$

The variation of link flow is given for every individual link. The travel time between two nodes is given by the summation of travel times of all links along the route.

Consideration of travel time reliability allows both the assumption that the variations in link flow are independent and that the variations in link flow are correlated. The characteristics and difference were compared and discussed (Wakabayashi and Iida, 1993).

When the distribution of travel time does not depend on the variation of traffic flow, and depends on the distribution of vehicles' speed, the probability distribution function $H_a(t)$ can be obtained directly.

5 WEATHER FORECAST AND NETWORK RELIABILITY

The traffic control on the expressway is carried out according to the actual weather, not to the weather forecast. The weather forecast seems to be used for the preparation of traffic control. In this study, the longitudinal data of traffic control could be obtained. The actual local weather, however, along the expressway cannot be known, excluding the weather monitoring points. Thus, only the relationship between traffic control and weather forecast could be used for the reliability analysis.

In this section, the outline of the weather forecast related to the traffic control is described. Japan Meteorological Agency (JMA) identifies "weather" as the comprehensive atmospheric condition including the phenomena caused by atmospheric vapors like the hydrometeor (rainfall) and cloud, those caused by electricity like the thunder as well as the basic physical

properties like the wind and temperature. JMA provides six kinds of weather forecast;
1) Weather forecast of today, tomorrow, and the day after tomorrow,
2) Weather distribution forecast,
3) Time series forecast,
4) Short period forecast of rainfall,
5) Weekly weather forecast, and;
6) Long range weather forecast.

Major weather forecasts are 1), 2) and 3) that are provided three times a day. 1) is a "today, tomorrow and the day after tomorrow weather forecast" that is very familiar to people, provided every 5 a.m., 11 a.m. and 5 p.m. This forecast consists of weather, wind, temperature, waves and rainfall probabilities. 2) is a weather distribution forecast which predicts weather, wind temperature and rainfall probability for every 3 hours for the next 24 hours for every 20 km squared mesh, provided every 6 a.m., noon, 6 p.m.. 3) is a time series forecast, which predicts weather and temperature at fixed points of every prefecture every 3 hours for the next 24 hours. JMA began to offer forecast of 2) and 3) in order to provide spatially and longitudinally more accurate weather forecast since March, 1996.

The rainfall probability is defined as the probability of rainfall or snowfall of more than 1 mm in six hours and is given as a round number between 0% and 100%. The hit ratio of the weather forecast is reported to be 81%.

However, since the time series recorded data for 2) and 3) are not available yet, 1) *i.e.* "weather forecasts of today, tomorrow and the day after tomorrow" is used in this study. The data used in this study related to traffic control in the expressway, are "weather" and "rainfall (snowfall) probability". Although, previous cumulate data of these weather forecasts are provided from every local JMA observatory, spatially aggregated weather forecast data are not provided. To obtain the aggregated data for the whole study area, we would need to visit all local JMA observatories and this is very time consuming. Instead of this, we chose to use some newspaper weather forecasts, that are only locally integrated.

6 SNOWFALL WEATHER FORECAST AND EXPRESSWAY NETWORK RELIABILITY ASSESSMENT

This section presents a brief case study of reliability assessment of expressway under snowfall environment. First, the traffic control under snowfall environment is stated. Second, the chosen network is described. Third, a very brief reliability analysis is carried out.

Types of Traffic Control under Snowfall Environment

In Japan Highway Public Corporation, five types of traffic regulation can apply: 1) “closed”, 2) “snowplow operation”, 3) 50 km/h speed regulation, 4) 80 km/h speed regulation, and 5) no regulation. “Snowplow operation” means that paralleled snowplows are located on each lane, removing snow at 50 km/h speed.

Chosen Network

On the first phase of this study, we focus on the existing expressway network. There are two routes between Nagoya and Osaka, that is, the Meishin Expressway Route and the Higashi-Meihan Expressway Route. The former consists of the end part of the Tohmei (Tokyo and Nagoya) Expressway and main body of the Meishin (Nagoya and Kobe) Expressway. The studied node pair is between Nagoya Interchange (I.C.) and Suita Junction (J.C.T.). The route length is 188.6 km. The latter consists of the Higashi-Meihan (eastern part of Nagoya and Osaka) Expressway, National Route No.25, Nishi-Meihan Expressway (western part of Nagoya and Osaka), and Kinki Expressway. The same node pair is studied.

Network Reliability Assessment for Normal Period

Here, the events for network availability are traffic accidents. The average accident rates are given as 234 accidents / 10 million traveled kilometer vehicle for wet road surfaces, and 31 accidents / 10 million traveled kilometer vehicle for dry road surfaces.

Assuming the Poisson occurrence of accidents, the probability p_x that the number of accidents in link a with link length L_a is equal to x is given by

$$p_x = (\lambda L_a)^x e^{-\lambda L_a} / x! \ . \tag{31}$$

where λ is average accident occurrence rate per kilometer.
From Eq. (31), the probability of no accidents occurring per unit link length is given by

$$r_{aN} = \exp(-\lambda L_a) . \tag{32}$$

Since Poisson occurrence is considered, no consideration for dependent events is required. The

value for λ is taken to be the average in wet and dry conditions.
From the accident rates and link flow data, a set of link reliabilities is calculated for all links. Node to node reliability was calculated as 0.982 for the Meishin route and as 0.967 for the Meihan route. The reliability of connection for a normal period was then calculated as 0.999.

Network Reliability Assessment under Snowfall Environment

Two types of conditional probability are introduced. One is based on “weather forecast of snowfall”. The other is based on “weather forecast with snowfall probability”. The weather forecast database is constructed from various newspapers’ data sources. From this database, the following link reliabilities are calculated.

The first type of link reliability is based on “weather forecast of snowfall”. The link reliability is given by

$$r_{aS1} = 1 - \Pr\{\text{link } a \text{ is closed} \mid \text{weather forecast is "snow"}\}. \tag{33}$$

Based on this reliability and without considering inter-dependent events, the node to node reliability is calculated as 0.091 for the Meishin route and as 0.065 for the Meihan route.

The second type of link reliability is based on “snowfall probability”. The link reliability is then given by

$$r_{aS2} = 1 - \Pr\{\text{link } a \text{ is closed} \mid \text{rank } i \text{ of snowfall probability}\}. \tag{34}$$

where rank 1,2 and 3 indicate {0,10,20%}, {30,40,50%}, {60,70,80,90,100%} of snowfall probability.

7 CONCLUSION

This chapter presents a methodology for assessing highway network reliability in a snowfall environment. First, the methods for assessing system reliability proposed by many researchers were discussed, and were classified into several categories. However, both the required computational work and the memory size for each method increase exponentially with the size of the network. This is the most difficult point in assessing system reliability. Second, two methods proposed by the authors are presented. They are efficient methods as they limit the employed number of minimal paths and cuts. Third, the necessity for including dependent events consideration was discussed. Two approaches for considering dependent events are set out. One is an approximate method using a correlation indicator. The methodology and numerical example are reported. Another is a method using travel time reliability. Lastly, weather forecast and network reliability are discussed. After introducing the type of weather forecast provided in Japan, the application of the reliability assessment method was carried out for an actual expressway network. The normal period considering the occurrence of traffic accidents and abnormal period in snowfall environment were compared.

8 REFERENCES

Barlow, R. E. and Proschan, F. (1965). Mathematical theory of reliability, John Wiley & Sons, Inc., New York.

Bell, M.G.H. and Iida, Y. (1997). Transportation network analysis, Wiley, 179-192.

Fratta, L. and Montanari, U.G., 1973. A Boolean algebra method for computing the terminal reliability in a communication network, IEEE Transaction on Circuit Theory, Vol. CT-20, **3**, 203-211.

Henley, E. J. and Kumamoto, H. (1981). Reliability engineering and risk assessment, Prentice-Hall, Inc..

Wakabayashi,H. and Iida,Y. (1992). Upper and lower bounds of terminal reliability of road networks: an efficient method with Boolean algebra. ***J.*** of Natural Disaster Science, Vol.14, **1**, 29-44.

Wakabayashi,H. and Iida,Y. (1993). Improvement of terminal reliability and travel time reliability under traffic management. Pacific Rim TransTech Conference Proceedings Volume 1, Advanced Technologies, ASCE 3rd International Conference on Applications of Advanced Technologies in Transportation Engineering, ASCE, 211-217.

Wakabayashi,H. and Iida,Y. (1994). Improvement of road network reliability with traffic management. In: B.Liu and J.M.Blosseville (eds.), Transportation Systems: Theory and Applications of Advanced Technology. Pergamon Elsevier Science, IFAC, 603-608.

CHAPTER 9

TRAVEL TIME VERSUS CAPACITY RELIABILITY OF A ROAD NETWORK

Hai Yang, Ka Kan Lo and Wilson H. Tang

Department of Civil Engineering, The Hong Kong University of Science and Technology, Clear Water Bay, Kowloon, Hong Kong, P.R. China

INTRODUCTION

Travel time reliability and capacity reliability are considered to be two useful measures of the performance of a road traffic network. The former is concerned with the probability that a trip between a given origin-destination pair can be made successfully within a specified interval of time for a given level of traffic demand in the network; while the latter is given as the probability that the network can accommodate a certain traffic demand at a required level of service. Determination of both reliabilities requires accounting for drivers' route choice behavior. This chapter proposes to evaluate and synthesize the two interdependent reliabilities of a road network subject to link capacity degradation. A two-dimensional graphical approach is demonstrated to allow for intuitive perception of the changes of the travel time and capacity reliability at varied levels of traffic demand and service, thereby leading to a practical and useful reliability-based comprehensive performance measure of a road network.

1. OUTLINE OF PREVIOUS STUDIES

The economy in a nation or region depends heavily upon an efficient and reliable transportation system to provide accessibility and promote the safe and efficient movement of people and goods. Moreover, a reliable transportation system is essential for maintenance and repair of other lifeline systems. Thus the importance of guaranteeing an acceptable level of transportation service cannot be overemphasized. However, in reality smooth traffic flow in a road network is often subject to interruptions by events such as earthquakes, floods, traffic accidents, adverse weather and slope failure, for periods varying from a few hours to a few years. Thus it is necessary and important to study the reliability of a road traffic network.

1.1 Previous studies of the problem

With increasing needs for better and more reliable services, many systems (e.g., electrical power systems, water distribution system and communication networks) have incorporated reliability analysis as an integral part in their planning, design, and operation (Ang and Tang, 1990). However, reliability analysis has received very limited attention in the context of road traffic networks in spite of its importance. A few existing reliability studies of road networks are mainly limited to three aspects: connectivity, travel time reliability and capacity reliability (Wakabayashi and Iida, 1992; Bell and Iida, 1997, Chen *et al.*, 1999a & 1999b). A general theoretical framework and a review for reliability analysis of degradable transportation systems can be found in Du and Nicholson (1997).

Connectivity reliability. Connectivity reliability is concerned with the probability that the network nodes remain connected. A special case of the connectivity reliability is the terminal reliability, which concerns the existence of a path between a specific origin-destination (O-D) pair (Iida and Wakabayashi, 1989). For each node pair, the network is considered successful if at least one path is operational. A path consists of a set of components (roadways or links) which are characterized by a zero-one variable denoting the state of each link (operating or failed). Capacity constraints on the links are not accounted for when finding the connectivity reliability. This type of reliability analysis might be suitable for abnormal situations, such as earthquakes, but there is an inherent deficiency in the sense that it only allows for two operating states: operating at full capacity or complete failure with zero capacity. This binary state approach limits the application to everyday situations where links are operating in-between these two extremes. The approach may thus result in misleading results of reliability and risk assessment of a road network.

Travel time reliability. Travel time reliability is introduced to reflect variation of O-D travel time (Asakura and Kashiwadani, 1991). It is concerned with the probability that a trip between a given O-D pair can be made successfully within a specified interval of time. Bell *et al.* (1998) proposed a sensitivity analysis based procedure to estimate the variance of travel time arising from daily demand fluctuations. Asakura (1996) examined travel time reliability in the case of capacity degradation due to deteriorated roads. Travel time reliability is given as a function of the ratio of travel times under the degraded and non-degraded states. When the ratio is close to unity, it is essentially operating at ideal capacity; whereas when it approaches infinity, the destination is not reachable because certain links are so severely degraded. This extreme case is consistent with network connectivity reliability. Generally, the measure of travel time reliability is useful to evaluate network performance in terms of service quality that should be maintained in daily operations.

Capacity reliability. Generally speaking, the aforementioned connectivity and travel time reliability measures are useful for assessing the quality of service that is of interest to individual drivers. The capacity reliability of the network - the probability that the network capacity can successfully accommodate a certain level of O-D demand at an acceptable service quality - should be considered an important and meaningful measure of overall system performance that is of interest to system managers. The reason is simple: the connectivity between the origin and the destination is a necessary condition for the successful operation of a road network, but it is not a sufficient condition. The success of the O-D connection should also ensure the availability of the required O-D capacity. Thus capacity reliability should be considered together with travel time reliability. Efforts to incorporate capacity into reliability have been made for communication networks (Aggarwal, 1985; Chan *el al.*, 1997), water distribution systems (Li *et al.*, 1993), and electric power systems (Billington and Li, 1994) to determine the maximum flow of the networks. However, the approach is not directly applicable to a road traffic network where the capacity modeling characteristics is quite different in the following ways:

- The movement of flows in the road network involves flows of people rather than pure physical commodities treated in classic network flow theory, and the travel delay will increase with increasing flow as a result of congestion. Thus drivers' route choice behavior has to be considered in determining maximum flow of a congested road network.
- Conventionally, the capacity of a physical commodity network is defined as the largest possible sum of flow from a source to a sink capable of being accommodated, while honoring a given capacity on each link. However, road network capacity or maximum flow level should be specified with regard to its level of service such as O-D travel time.
- In addition, multiple O-D pairs exist and the flow is between different O-D pairs are not exchangeable or substitutable in modeling road network capacity. It is thus important to define the O-D demand pattern that greatly influences the resultant value of road network capacity. It is important to select the target O-D matrix appropriately for capacity calculation.

The above characteristics make the modeling of road network travel time and capacity a quite complex, yet intriguing problem. Recently, Chen *et al.* (1999a & 1999b) conducted an extensive simulation study of network capacity reliability and sensitivity analysis using the concept of reserve capacity of a road network proposed by Wong and Yang (1997).

1.2 The objectives and organization of the study

As mentioned above, both travel time and capacity reliability are important and useful measures of the performance of a road network. The two reliabilities are essentially interdependent for a given network, but so far they have been proposed and modeled separately. It is intriguing to investigate their relationship and how the two complementary reliability measures can be synthesized. For instance, for a given network subject to recurrent or non-recurrent capacity disturbances, travel time or level of service will be different at different levels of demand due to different degrees of flow variation and congestion. However, maximum network flow or capacity will vary at different required levels of service. Thus while a single measure of either travel time or capacity reliability might not be sufficient, their combination may prove to be useful as a comprehensive performance measure of a road network. Therefore, it would be interesting and meaningful to develop methods for the evaluation and integration of road network travel time and capacity reliability. This will constitute a more useful comprehensive performance measure of a road network. This performance measure can be used for robust network planning and design.

The objective of the proposed study is to develop methods for the evaluation and synthesis of both travel time and capacity reliability of a road network, thereby forming a comprehensive network performance measure. The next section introduces both travel time and capacity reliability of a road network subject to link capacity variations. Section 3 investigates the relationship between the capacity and travel time reliability and shows how the two concepts can be synthesized to form a more comprehensive network performance measure. A simple analytical example is provided to demonstrate how a two-dimensional graphical approach can be used to synthesize the two reliabilities. Conclusions are presented in Section 4.

2. TRAVEL TIME AND CAPACITY RELIABILITY

2.1 Route choice behavior of drivers and road capacity variations

Consider a road network $G(N, A)$ where N is the set of network nodes and A is the set of network links. Congestion effect is considered by using flow-dependent link travel time function $t_a(v_a, c_a), a \in A$ where v_a is the flow on link $a \in A$ and c_a is its capacity. Drivers' route choices are modeled by the following standard user-equilibrium traffic assignment model (e.g., Patriksson, 1994):

$$\underset{v}{\text{minimize}} \sum_{a \in A} \int_0^{v_a} t_a(\omega, c_a) d\omega \tag{1}$$

subject to:

$$\sum_{r \in R_w} f_w^r = q_w, w \in W \tag{2}$$

$$v_a = \sum_{w \in W} \sum_{r \in R_w} f_w^r \delta_{ar}^w, \; a \in A \tag{3}$$

$$f_w^r \geq 0, r \in R_w, w \in W \tag{4}$$

where R is the set of routes in the network and R_w is the set of routes between O-D pair $w \in W$ and W is the set of O-D pairs; q_w is the demand between O-D pair $w \in W$ and f_w^r is the flow on route $r \in R$; δ_{ar}^w equals 1 if route r between O-D pair w uses link a and 0 otherwise. The optimum solution $\mathbf{f}^* = (\cdots, f_w^r, \cdots)$ satisfies the following user-equilibrium conditions:

$$u_w^r(\mathbf{f}^*) - u_w(\mathbf{f}^*) \begin{cases} = 0 & \text{if } f_w^r > 0, \\ \geq 0 & \text{if } f_w^r = 0. \end{cases} \tag{5}$$

where $u_w^r(\mathbf{f}^*) = \sum_{a \in A} t_a(v_a^*)\delta_{ar}^w$ is the travel time on path $r \in R_w$, $u_w(\mathbf{f}^*) = \min(u_w^r(\mathbf{f}^*), r \in R_w)$ is the minimum travel time between O-D pair $w \in W$. That is, when the travel time on path r is larger than or equal to the shortest travel time, the flow on that path is zero or the path is not used. When the travel time on path r is equal to the minimum one, its flow is greater than or equal to zero.

Clearly, the resultant network travel time and maximum network flow depends on the link capacity: $\mathbf{c} = (\cdots, c_a, \cdots)$. In reality, road capacity is not deterministic, but subject to variations due to traffic accidents, weather conditions, landslides and so on. Let $\mathbf{c} = \mathbf{c}_0 - \varepsilon$, where $\mathbf{c}_0 = (\cdots, c_{a0}, \cdots)$ is a vector of normal link capacities and ε is a vector of random variables representing link capacity degradation. Variable ε may range from 0 (no degradation) to $\mathbf{c}_0$ with complete degradation (or zero capacity). We suppose the probability distribution $p(\varepsilon)$ of occurrence of the perturbation vector ε is available, which can be estimated from existing sources of data.

Given the fact that link capacities are random variables following a certain probability distribution, we are interested in knowing the resulting probabilistic fluctuations or reliabilities of travel time and maximum network flow through the network.

2.2 Travel time reliability of a road network

Travel time is uncertain due to degradable road components and its reliability can be defined as the probability that a trip between an O-D pair can be successfully made within a specified interval of time. Because travel time depends on the degree of congestion which, in turn, depends on travel demand on the network, its reliability should be defined with respect to a given reference O-D demand pattern. Let $\mathbf{q}_0 = (\cdots, q_{w0}, \cdots)$ be a vector of a given basic reference O-D demand pattern (e.g., the current level of O-D demands). To investigate travel time reliability at variable levels of demand, we introduce a factor $\tilde{\mu}$ to scale up or down the basic O-D demand pattern uniformly, $\mathbf{q} = \tilde{\mu}\mathbf{q}_0$ and determine the resulting travel time reliability. Let $u_w(\mathbf{c}_0, \mathbf{q}_0)$ be the travel time between O-D pair $w \in W$ corresponding to the basic reference O-D demand $\mathbf{q}_0$ and normal link capacities in a non-degraded state, and let $u_w(\mathbf{c}, \mathbf{q})$ be the travel time between O-D pair $w \in W$ corresponding to the multiplied reference O-D demand $\tilde{\mu}\mathbf{q}_0$ and link capacities in a degraded state. Clearly, $u_w(\mathbf{c}, \mathbf{q})$ is a random variable because of random variation of capacity $\mathbf{c}$. The travel time reliability under the multiplied O-D demand for an individual O-D pair $w \in W$ is denoted as $TR_w(\tilde{\pi}, \tilde{\mu})$ and defined as the probability of whether the ratio of $u_w(\mathbf{c}, \mathbf{q})$ to $u_w(\mathbf{c}_0, \mathbf{q}_0)$ is kept within an acceptable level of threshold $\tilde{\pi}$. Namely,

$$TR_w(\tilde{\pi}, \tilde{\mu}) = \Pr\left\{ \frac{u_w(\mathbf{c}, \mathbf{q})}{u_w(\mathbf{c}_0, \mathbf{q}_0)} \leq \tilde{\pi} \,\middle|\, \mathbf{q} = \tilde{\mu}\mathbf{q}_0 \right\},\ w \in W \quad (6)$$

Clearly, this probability predicts how reliably trips between an O-D pair can be made through a network with degradable links at a certain level of demand. The value $\tilde{\pi}$ can be interpreted as the level of service or travel time threshold that should be maintained.

Note that equation (6) is defined with respect to individual O-D pairs and the resulting reliability may vary across O-D pairs. It is sometimes more convenient to use a single index to describe the overall performance of the road network. In this case, the overall travel time reliability of the network can be defined as the probability that the travel time ratio between every O-D pair under degraded and normal conditions is less than a common threshold:

$$TR(\tilde{\pi},\tilde{\mu}) = \Pr\left\{ \frac{u_w(\mathbf{c},\mathbf{q})}{u_w(\mathbf{c}_0,\mathbf{q}_0)} \le \tilde{\pi},\ \forall w \in W \middle| \mathbf{q} = \tilde{\mu}\mathbf{q}_0 \right\} \tag{7}$$

where $TR(\tilde{\pi},\tilde{\mu})$ is the overall network travel time reliability for prescribed O-D demand level $\tilde{\pi}$ and travel time threshold $\tilde{\mu}$. Note that this might be a conservative measure with a strong requirement and might result in underestimation of the overall network travel time reliability. Nevertheless it can be relaxed by adopting an arithmetic average or demand-weighted average of all O-D travel time reliabilities.

2.3 Capacity reliability of a road network

Now we examine the maximum service flow rate and its reliability in a road network. We define the maximum service flow rate as the maximum network flow throughput at a prescribed level of service. For consistence and comparison with the travel time reliability, the level of service is specified in such a way that travel time $u_w(\mathbf{c},\mathbf{q})$ in a degradable state between each O-D pair should not exceed $\tilde{\pi}$ times the basic reference travel time $u_w(\mathbf{c}_0,\mathbf{q}_0)$. Chen et al. (1999a, 1999b) defined the capacity reliability of a network as the probability that the network reserve capacity can accommodate a certain traffic demand at the prescribed service level. Reserve capacity is defined as the largest multiplier μ applied to the given basic O-D demand $\mathbf{q}_0$ that can be allocated to a network without violating the link capacities. Here this definition is slightly modified by adding one additional constraint that the O-D travel time resulting from the equilibrium traffic assignment (1)-(4) does not exceed the corresponding prescribed travel time threshold:

$$\max \mu \text{, s.t. } u_w(\mathbf{c},\mathbf{q}) \le \tilde{\pi} u_w(\mathbf{c}_0,\mathbf{q}_0),\ q_w = \mu q_{w0},\ w \in W,\ v_a(\mathbf{c},\mathbf{q}) \le c_a,\ a \in A \tag{8}$$

Clearly, the largest O-D matrix multiplier μ depends on link capacities $\mathbf{c}$. Instead of finding the maximum μ with deterministic link capacities $\mathbf{c}_0$, we are interested in the probability that the network reserve capacity $\mu(\mathbf{c})$ with degradable link capacities $\mathbf{c}$ is greater than or equal to the required demand level $\mathbf{q}$ specified by a predetermined demand multiplier $\tilde{\mu}$ $(\mathbf{q} = \tilde{\mu}\mathbf{q}_0)$, when link capacity is subject to random variations. This probability is defined as the overall network capacity reliability $CR(\tilde{\pi},\tilde{\mu})$ and is given below.

$$CR(\widetilde{\pi},\widetilde{\mu}) = \Pr\left\{\exists r, v(a)\delta_{ar}^{w} \leq c_a, \forall a, \forall w \left| \frac{u_w(\mathbf{c},\mathbf{q})}{u_w(\mathbf{c}_0,\mathbf{q}_0)} \leq \widetilde{\pi}, \forall w \in W, \mathbf{q} = \widetilde{\mu}\mathbf{q}_0 \right.\right\} \quad (9)$$

Note that this definition requires that at least one unsaturated path (a path not including over-saturated links) is available between each O-D pair.

In reality an O-D pair may be operational even when some other O-D pairs are over-saturated or blocked. We thus introduce O-D pair specific capacity reliability $CR_w(\widetilde{\pi},\widetilde{\mu})$ as follows:

$$CR_w(\widetilde{\pi},\widetilde{\mu}) = \Pr\left\{\exists r, v(a)\delta_{ar}^{w} \leq c_a, \forall a \left| \frac{u_w(\mathbf{c},\mathbf{q})}{u_w(\mathbf{c}_0,\mathbf{q}_0)} \leq \widetilde{\pi}, \mathbf{q} = \widetilde{\mu}\mathbf{q}_0 \right.\right\}, \ w \in W \quad (10)$$

Thus by varying the values of $\widetilde{\pi}$ and $\widetilde{\mu}$, we can predict how reliably a network with degradable links can accommodate various required levels of demand at different required levels of service in terms of a travel time threshold.

3. SYNTHESIS OF TRAVEL TIME AND CAPACITY RELIABILITY

3.1 Interdependence of travel time and capacity reliability

It is interesting to see that the aforementioned travel time reliability $TR(\widetilde{\pi},\widetilde{\mu})$ or $TR_w(\widetilde{\pi},\widetilde{\mu})$ shown in (7) and (6) and capacity reliability $CR(\widetilde{\pi},\widetilde{\mu})$ or $CR_w(\widetilde{\pi},\widetilde{\mu})$ in (9) and (10) are intertwined through two intervening reference parameters: travel time threshold $\widetilde{\pi}$ and traffic demand threshold $\widetilde{\mu}$. This is not surprising if we recall the definitions of the two reliabilities. Travel time reliability is given as the probability that a trip between an O-D pair can be successfully made within a specific interval of time. Travel time depends on the level of demand due to congestion and thus travel time reliability has to be specified with respect to a specific level of traffic demand. On the other hand, capacity reliability is defined as the probability that the network reserve capacity is greater than or equal to a required demand level at a minimum required level of service. This required level of service could be selected as a threshold of travel time (or scheduled travel time) between each O-D pair. Therefore, the two reliabilities for a given network are interdependent. Figure 1 shows diagrammatically the essential elements of these relationships.

The interdependence of travel time and capacity reliability makes it intriguing to investigate how the two complementary reliabilities will change against varied levels of threshold of travel time and traffic demand in the network, and integrate them to form a new comprehensive network performance index. Here we demonstrate a two-dimensional graphical approach to synthesis of the two reliabilities with a numerical example.

3.2 Numerical Simulation Settings

In order to show how to intertwine the two reliabilities, a simple network is tested. Figure 2 displays the simplified Tuen Mun Highway network in Hong Kong adopted here. There are in total four nodes and ten links with the following BPR functions as their link performance functions:

$$t_a = \alpha_a + \beta_a \left(\frac{v_a}{C_a} \right)^{\gamma_a} \tag{11}$$

where v_a, α_a, and C_a are the flow, free-flow travel time, and capacity of link *a*, respectively. β_a and γ_a are the calibrated parameters of link *a*. The values of the link characteristics (free-flow travel time and link capacity) from actual observations are given in Table 1. Table 2 provides the basic reference O-D demands.

Using the Monte Carlo simulation method, we develop a procedure to estimate the capacity reliability and travel time reliability distribution under different OD demand factors, levels of service and also link capacities. The procedure is shown in Figure 3. In our simulation we assume that capacity degradation of each network link is independent and the capacity reduction simply follows uniform distribution between normal and half-normal capacity. For each simulation run, only 25% of network links are subject to degradation.

3.3 Simulation results and analysis

Figures 4a and 4b depict the contours of individual O-D based (with 1→3 as an example) and network based capacity reliability respectively, where the vertical axis is the travel time threshold $\tilde{\pi}$ while the horizontal axis is the traffic demand threshold $\tilde{\mu}$. We can see that the contours of both figures are rationally distributed as we expect (note that curve smoothing manipulation is made for the data generated from the simulation). At a particular level of traffic demand, higher capacity reliability can be achieved only at the expense of lower level of service (i.e. higher travel time threshold). On the other hand, higher traffic demand results in lower capacity reliability at the same level of service. This is obvious due to the congestion effect on some links. In addition, from the two figures, we can observe that both O-D based and network-based capacity reliability contours exhibit very similar distribution patterns.

Figures 5a and 5b portray the contours of O-D based and network based travel time reliability, respectively. Clearly the contours are reasonably distributed. At a particular level of traffic demand, lower level of service (i.e. larger travel time threshold) corresponds to higher travel time reliability; lower traffic demand results in higher travel time reliability at the same level of service. In addition, by comparing Figures 5a and 5b with Figures 4a and 4b, we can see that travel time reliability contours (either O-D based or network based) are very close to the counterpart of capacity reliabilities. This closeness or difference may well depend on network structure and O-D demand pattern and many other factors.

Once the contours of the two types of reliabilities are established, a number of potential applications can be considered. Such examples of applications are shown in Figures 6a and 6b. These figures show the domain of traffic demand and travel time that can be achieved at the required travel time and capacity reliability, either for an individual O-D pair or for the whole network. The boundary of the domain provides the trade-off between maximum traffic demand and minimum travel time satisfying the same capacity and travel time reliabilities. Similarly, we can easily identify the travel time and/or capacity reliability that can be achieved for given traffic demand and level of service.

4. CONCLUSIONS

Travel time reliability and capacity reliability are studied and integrated as a useful comprehensive performance measure of a road network for a wide variety of traffic flow conditions. For a given network subject to link capacity variations, both reliabilities are investigated in a two-dimensional traffic demand and service level space and represented by iso-reliability contours. Our graphical representation allows for intuitive perception of how, and under what circumstances, a required level of travel time and capacity reliability can be achieved. Therefore, our modeling approach established an efficient procedure for a reliability-based comprehensive performance assessment of a road network.

The proposed approach can be extended to incorporate reliability in to network design problems (Yang and Bell, 1998): an issue concerning the optimal expansion of roadway capacities or addition of new links to a network to satisfy requirement for both travel time and capacity reliabilities. In this context, one possible formulation is to maximize the network reserve capacity subject to meeting a pre-specified service standard (e.g., mean travel time) with certain reliability (e.g., the probability of the mean time not exceeding a pre-specified threshold).

ACKNOWLEDGMENTS

The authors thank Anthony Chen of University of California at Irvine for his participation in the early stage of our reliability studies. Thanks also go to Qiang Meng of Hong Kong University of Science and Technology for his helpful comments. This research is supported by the Hong Kong Research Grants Council through a RGC-CERG grant (6208/99E).

REFERENCES

Ang, A.H.S. and Tang, W.H. (1990) *Probability concepts in engineering planning and design*: Vol.1. *Basic principles*, Vol.2. *Decision, risk and reliability*. John Wiley & Sons, New York.

Aggarwal, K.K. (1985) Integration of reliability and capacity in performance measure of a telecommunication network. *IEEE Transactions on Reliability* **34**, 184-186.

Asakura, Y. and Kashiwadani, M. (1991) Road network reliability caused by daily fluctuation of traffic flow. *Proceedings of the 19th PTRC Summer Annual Meeting*, Brighton, pp.73-84.

Asakura, Y. (1996) Reliability measures of an origin and destination pair in a deteriorated road network with variable flows. In: *Proceeding of the 4th Meeting of the EURO Working Group in Transportation*.

Bell, M.G.H. and Iida, Y. (1997) *Transportation network analysis*, Chapter 8: Network reliability (pp.179-192). John Wiley & Sons, West Sussex, England.

Bell, M.G.H., Cassir, C., Iida, Y. and Lam, W.H.K. (1998) A sensitivity based approach to reliability assessment. *Paper submitted to the 14th ISTTT*, July 20-23, 1999.

Billington, R. and Li, W. (1994) *Reliability assessment of electric power systems using Monte Carlo methods*. Plenum Press, New York.

Chan, Y. Yim, E., and Marsh, A. (1997) Exact & approximate improvement to the throughput of a stochastic network. *IEEE Transactions on Reliability* **46**, 473-486.

Chen, A., Yang, H. and Lo, H., Tang, H. (1998) A capacity related reliability for transportation networks. *Journal of Advanced Transportation* (accepted).

Chen, A., Yang, H., Lo, H. and Tang, W. (1999) Capacity reliability of a road network: An assessment methodology and numerical results. *Transportation Research* (accepted).

Du, Z.P. and Nicholson, A. (1997) Degradable transportation systems: sensitivity and reliability analysis. *Transportation Research* **31B**, 225-237.

Iida, Y. and Wakabayashi, H. (1989) An approximation method of terminal reliability of a road network using partial minimal path and cut set. *Proceedings of the 5th WCTR*, Yokohama, pp.367-380.

Li, D., Dolezal, T., and Haimes, Y. (1993) Capacity reliability of water distribution networks. *Reliability Engineering and System Safety* **42**, 29-38.

Patriksson, M. (1994). *The traffic assignment problem: models and methods*. VSP, Utrecht, The Netherlands.

Wakabayashi, H. and Iida, Y. (1992) Upper and lower bounds of terminal reliability of road networks: an efficient method with Boolean algebra. *Journal of Natural Disaster Science* **14**, 29-44.

Wong, S.C. and Yang, H. (1997) Reserve capacity of a signal-controlled road network. *Transportation Research* **31B**, 397-402.

Yang, H. and Bell, M.G.H. (1998) Models and algorithms for road network design: a review and some new developments. *Transport Review* **18**, 257-278.

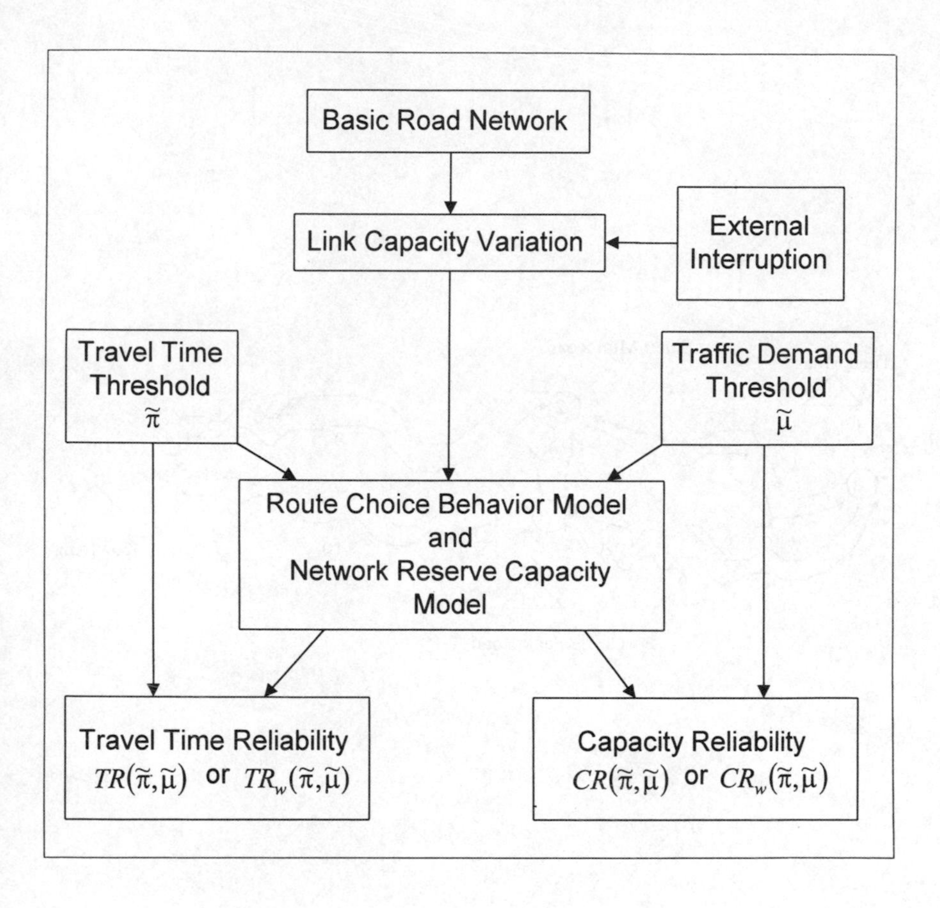

Figure 1. Interdependence of travel time and capacity reliabilities through two intervening variables: travel time and traffic demand thresholds.

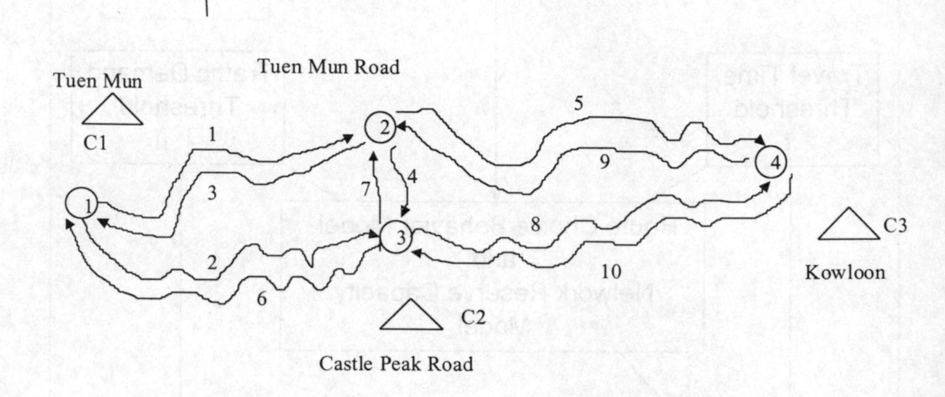

Figure 2: The simplified Tuen Mun Network.

Table 1: Link data of the Tuen Mun Road network.

Link #	T_f (hr)	Capacity (pcu/hr)	α	β
1	0.090	5175	0.1050	3.5
2	0.1106	850	0.1408	3.6
3	0.090	5175	0.1050	3.5
4	0.0056	730	0.0071	3.6
5	0.0335	4800	0.0335	3.6
6	0.1106	850	0.1408	3.6
7	0.0056	950	0.0071	3.6
8	0.0767	1000	0.1073	3.6
9	0.0335	4800	0.0335	3.6
10	0.0767	1000	0.1073	3.6

Table 2. Basic reference O-D matrix for simulation of the Tuen Mun network.

	1	3	4	Generation
1	-	110	2476	2586
3	156	-	63	219
4	2246	39	-	2285
Attraction	2402	149	2539	5090

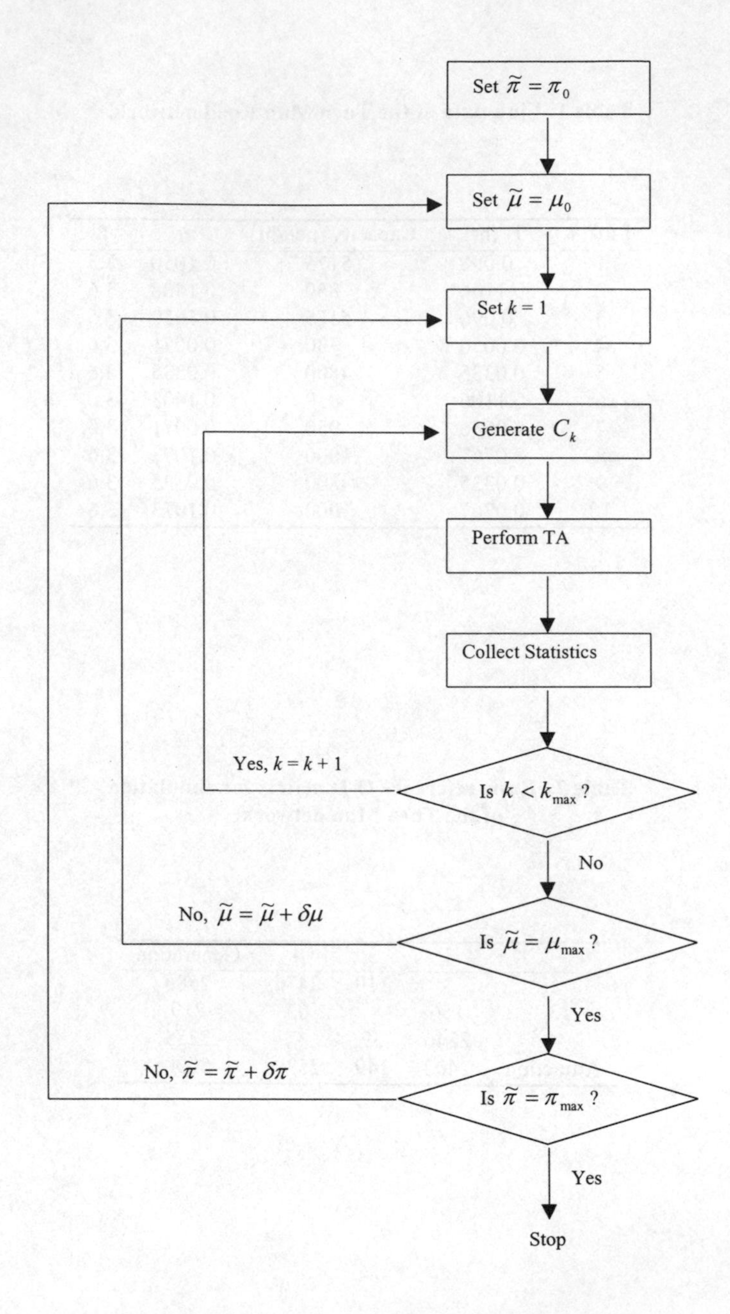

Figure 3: Flow chart of Monte Carlo simulation procedure.

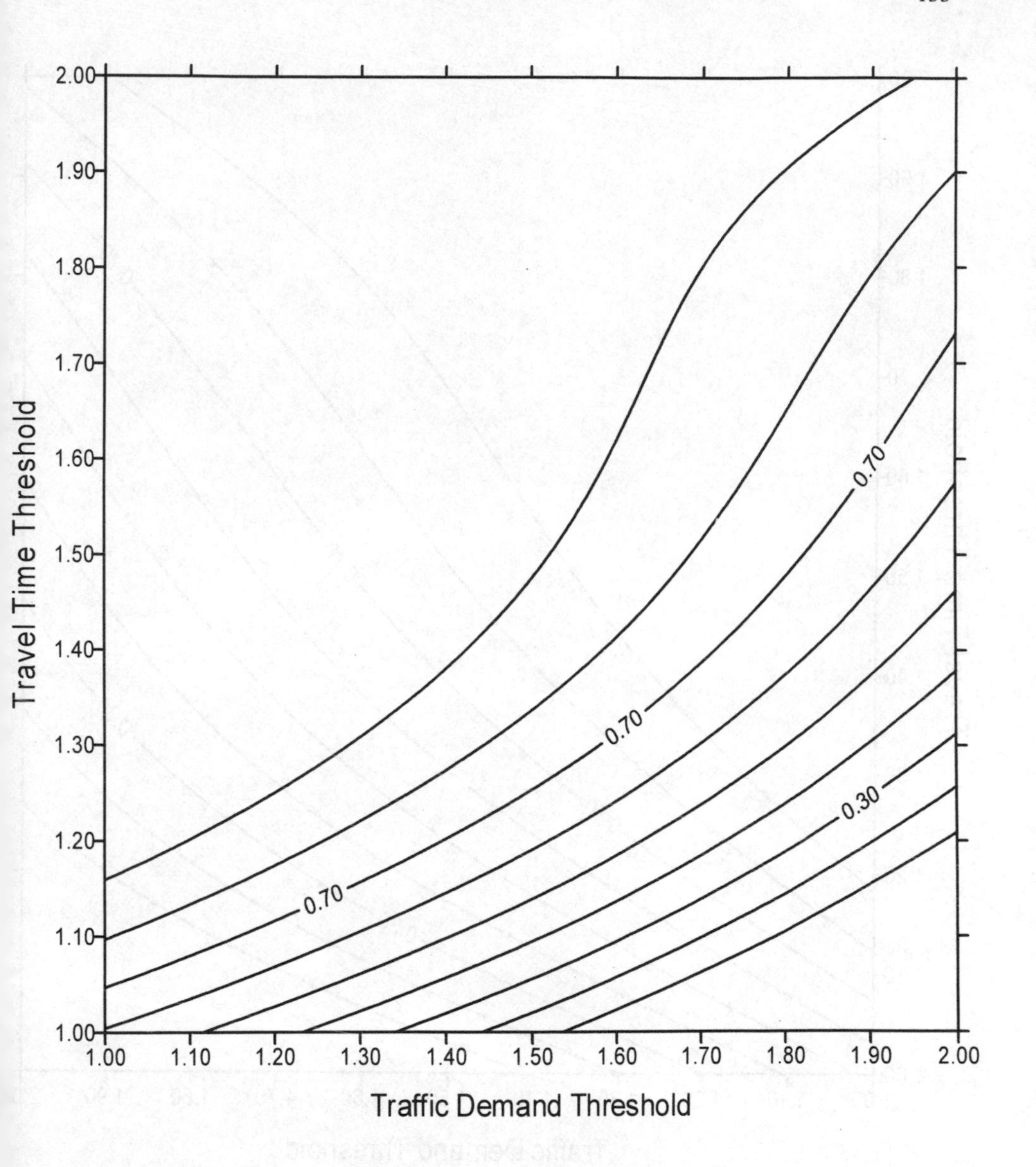

Figure 4a: Contours of individual O-D based capacity reliability

(origin 1 to destination 3).

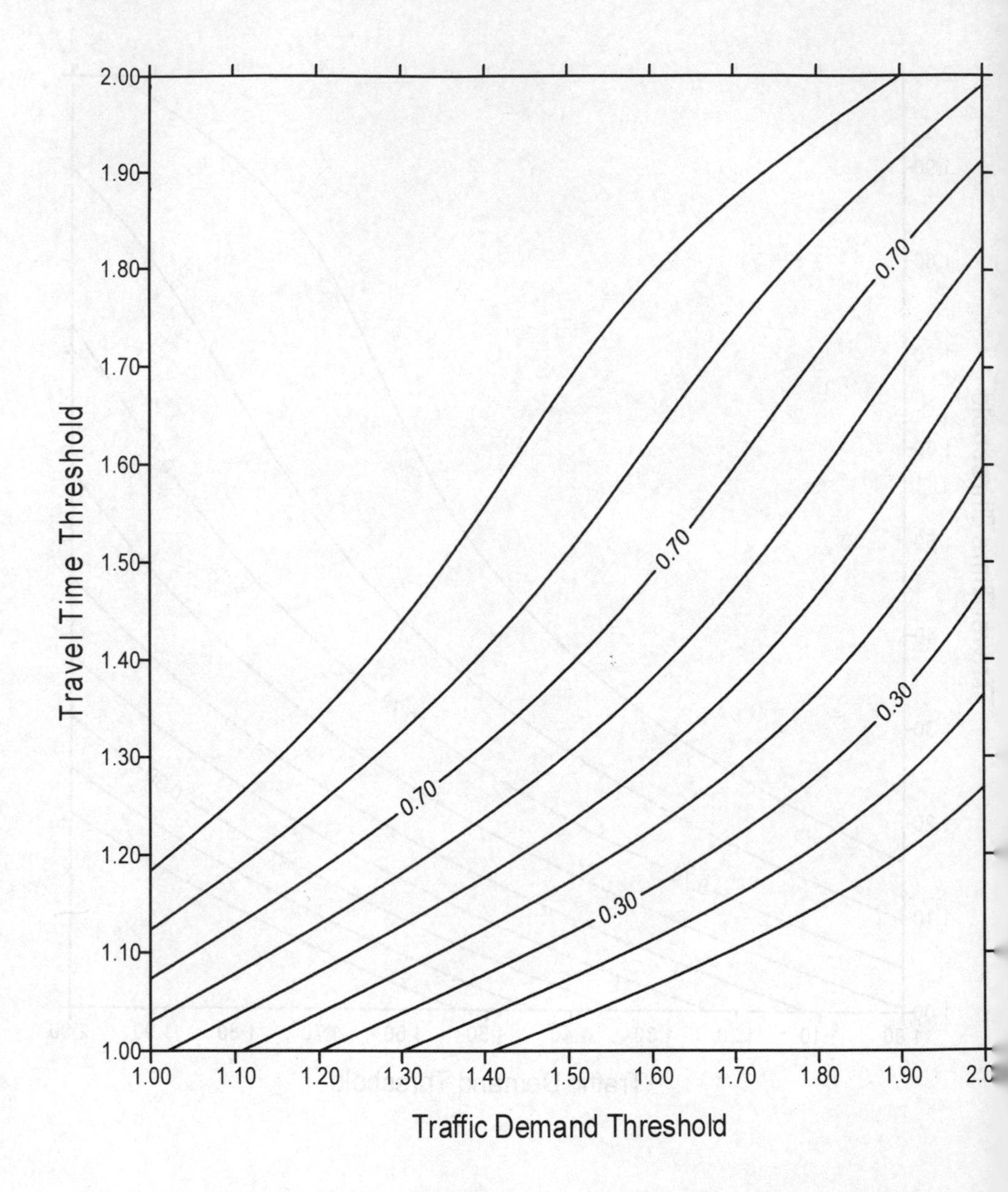

Figure 4b: Contours of overall network-based capacity reliability.

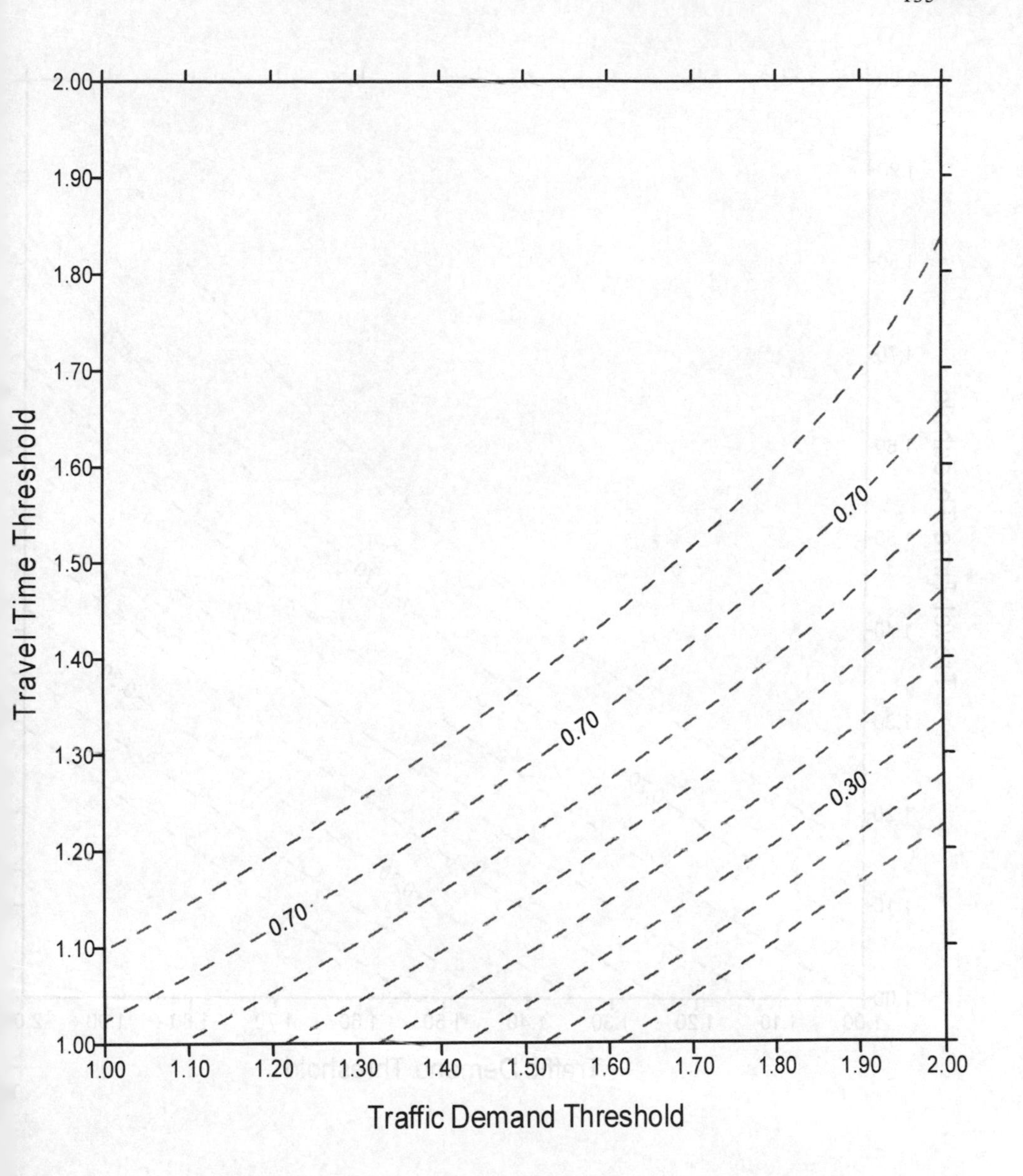

Figure 5a: Contours of individual O-D based travel time reliability

(origin 1 to destination 3).

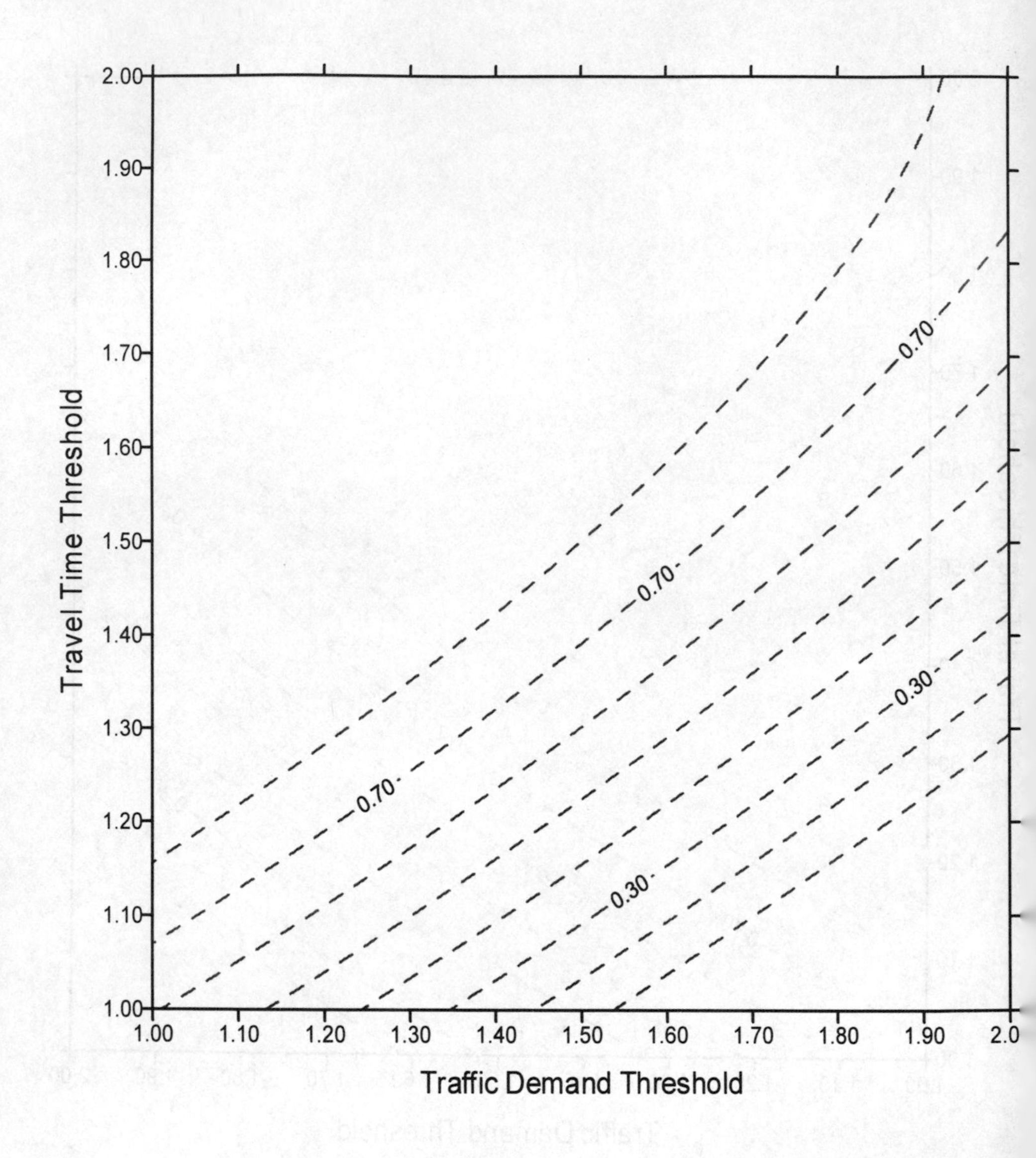

Figure 5b: Contours of overall network-based travel time reliability.

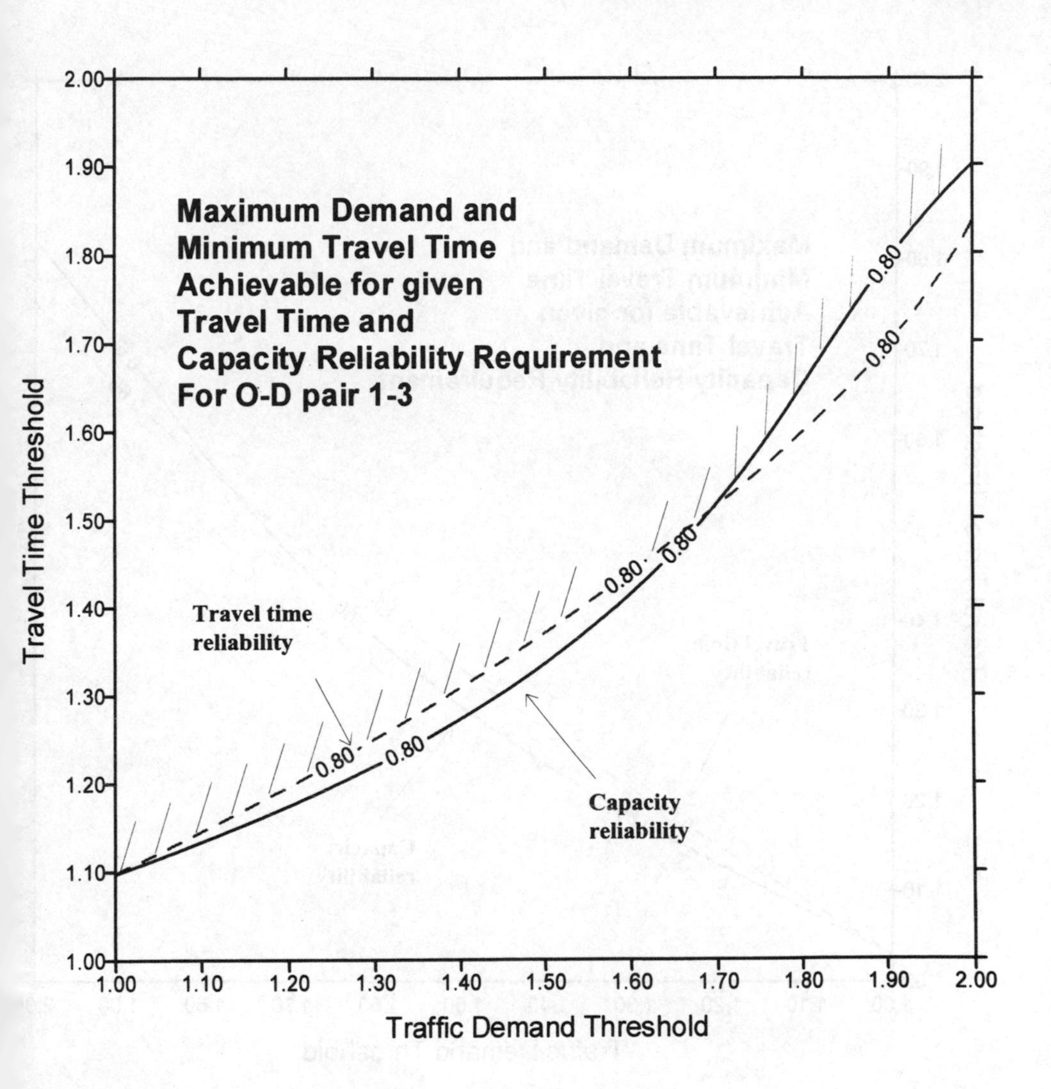

Figure 6a: Example for application of two O-D based reliability contours.

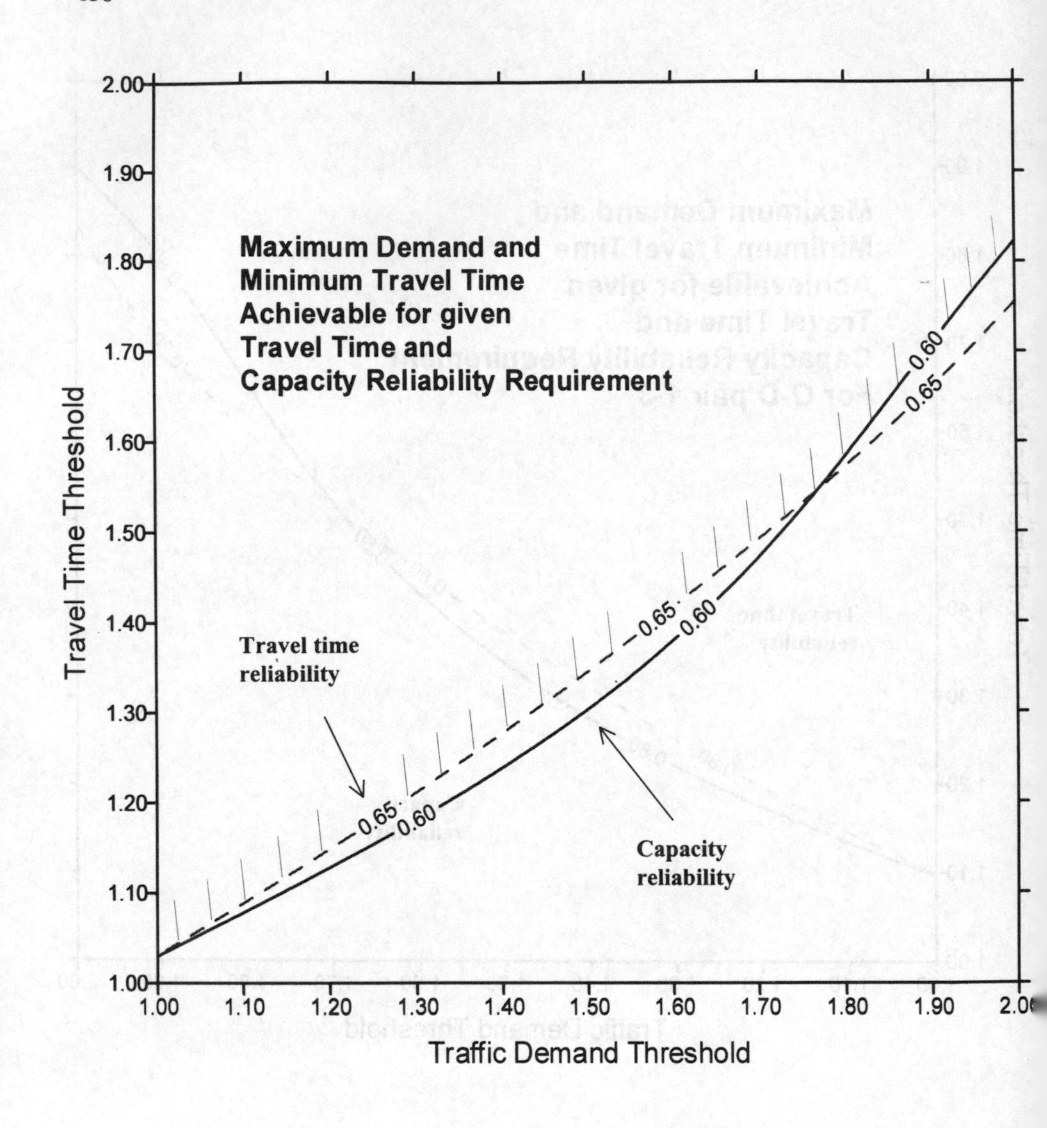

Figure 6b: Example for application of two network based reliability contours.

CHAPTER 10

CALIBRATION OF TRAFFIC FLOW SIMULATOR FOR NETWORK RELIABILITY ASSESSMENT

William H. K. Lam[1] *and G. Xu*

Department of Civil and Structural Engineering, The Hong Kong Polytechnic University, Hung Hom, Kowloon, Hong Kong Special Administration Region, China

INTRODUCTION

Traditional transportation network analysis can provide a complete picture of road traffic within the network on a static basis. However, the actual travel demands do vary over time, thereby contributing to the uncertainty of travel times. With the increased value of time, great loss is incurred by the drivers due to unexpected delay. Therefore more attention has recently been given to the analysis of transportation network reliability.

Network reliability can be assessed in two aspects, namely, connectivity reliability and travel time reliability (Bell and Iida, 1997; Iida, 1999). Connectivity reliability is the probability that there exists at least one path connected to the destination at worst. Travel time reliability indicates the stability of travel time, which is the probability of a trip to be completed within a given period of time. This can be affected by the fluctuating link flows due to origin-destination (OD) flow variation and the imperfect knowledge of the drivers when making route choice decision.

A traffic flow simulator (TFS) has recently been formulated by Lam and Xu (1999) to assess the transportation network reliability in terms of travel time reliability, which will be reviewed briefly below. The TFS is in fact a constrained probit-type stochastic user equilibrium (SUE) model with the assumption that drivers make their route choice decision based on their perceived path travel time that may be different from the actual travel time they experienced. The perception error across the drivers on a particular route (or path) is assumed to follow a normal distribution with zero mean and the variance is proportional to the actual path travel

[1] Author for correspondence: Fax (852) 2334-6389; Tel (852) 2766-6045; e-mail:cehklam@polyu.edu.hk

time. The ratio of the variance of the perception error to the actual path travel time is defined as perception error coefficient ω. On the other hand, Maher (1992) proposed a dispersion coefficient (i.e. the ratio of variance to mean for link travel times) for the SUE assignment model, in which a larger value taken by the dispersion parameter contributes to the increasing degree of multi-routing. Similarly, the assigned link flow pattern will be close to the user equilibrium (UE) solution when ω tends to zero. There is a need to calibrate the perception error coefficient (ω) before the application of TFS in practice. In addition, it is also required to calibrate the OD variation coefficient in the TFS.

The TFS calibration problem is complicated due to the complexity of the constrained probit-type SUE model and the effect of the OD variation. There are quite a number of studies devoted to the calibration of multinomial probit (MNP) models (Daganzo, 1979; Sparmann et al., 1983; Ghareib, 1996; Yai et al., 1997; Bolduc, 1999). The MNP model is a discrete choice model in which the attributes of individuals should be included in the survey data. The maximum likelihood (ML) method is the most common technique and has been widely used for the calibration of the MNP model based on a disaggregate data set. However, the kernel of the TFS is an aggregate probit model. Its calibration has to be based on an aggregate data set and therefore is different from the calibration of MNP models. In view of the calibration of aggregate models, there are also many efforts made for the calibration of logit-type choice models (Lam and Huang, 1992; Cascetta et al., 1996). But little attention has been given to the calibration of aggregate probit models, which is closely related to the calibration of TFS.

The genetic algorithm (GA) approach provides a new alternative for TFS calibration. The GA approach is advantaged by its capability for optimized stochastic search (Goldberg, 1989). It has been used widely for the optimization of complex systems. For instance, the GA technique has been used by Wong et al. (1998) for calibrating a land-use model (i.e. Lowry model). In this chapter, a calibration algorithm based on GA is proposed for the calibration of the TFS.

This chapter is structured as follows. Following the review of TFS, the measures that can be used for the TFS calibration are presented and discussed. Then the GA-based calibration approach is described. An example is employed for examination of the proposed calibration measures. Finally, conclusions are drawn together with recommendations for further study. Definitions of key variables that are employed in the remainder of the chapter are given in the Appendix.

REVIEW OF TFS

The TFS was proposed by Lam and Xu (1999) for the assessment of network travel time reliability based on the partial traffic counts and a prior OD matrix. It can be formulated as

$$\text{Min} \quad -\sum_{rs} q_{rs} S_{rs}[c^{rs}(v)] + \sum_a v_a t_a(v_a) - \sum_a \int_0^{v_a} t_a(w)dw + \lambda \sum_{rs} (q_{rs} - \hat{q}_{rs})^2 \tag{1}$$

$$\text{s.t.} \quad \sum_{rs} q_{rs} p_e^{rs} = \tilde{v}_e \text{, for links without detectors} \tag{2}$$

$$\sum_{rs} q_{rs} p_d^{rs} = \hat{v}_d \text{, for links with detectors} \tag{3}$$

$$(1-\delta)\hat{q}_{rs} \le q_{rs} \le (1+\delta)\hat{q}_{rs} \text{, for all OD pairs} \tag{4}$$

$$\tilde{\mathbf{v}}_\mathbf{e} = \bar{\mathbf{v}}_\mathbf{e} + \mathbf{B}_{21}\mathbf{B}_{11}^{-1}(\hat{\mathbf{v}}_\mathbf{d} - \bar{\mathbf{v}}_\mathbf{d}) \tag{5}$$

In TFS it is assumed that link flows are multivariate normally distributed random variables. Link flows can be denoted as $\mathbf{v} \sim MVN(\bar{\mathbf{v}}, \mathbf{B})$, where $\bar{\mathbf{v}}$ is the vector for the mean value of link flows and $\mathbf{B}$ is the variance/covariance matrix of link flow.

As partial traffic counts can be collected by detectors installed on some links, the link flows and variance/covariance matrix can be partitioned according to whether the links are installed with detectors or not. The order of the links can be rearranged so that the links with detectors appear first. If there are m out of n links installed with detectors, the link flow and mean link flow vector can be decomposed as shown below.

$$\mathbf{v} = [v_1 \quad \cdots \quad v_m \quad \vdots \quad v_{m+1} \quad \cdots \quad v_n]^T = [\mathbf{v}_\mathbf{d} \quad \mathbf{v}_\mathbf{e}]^T \tag{6}$$

$$\bar{\mathbf{v}} = [\bar{v}_1 \quad \cdots \quad \bar{v}_m \quad \vdots \quad \bar{v}_{m+1} \quad \cdots \quad \bar{v}_n]^T = [\bar{\mathbf{v}}_\mathbf{d} \quad \bar{\mathbf{v}}_\mathbf{e}]^T \tag{7}$$

Similarly, the link flow variance/covariance matrix $\mathbf{B}$ can also be decomposed as:

$$\mathbf{B} = \left[\begin{array}{ccc|ccc} b_{1,1} & \cdots & b_{1,m} & b_{1,m+1} & \cdots & b_{1,n} \\ \cdots & \cdots & \cdots & \cdots & \cdots & \cdots \\ b_{m,1} & \cdots & b_{m,m} & b_{m,m+1} & \cdots & b_{m,n} \\ \hline b_{m+1,1} & \cdots & b_{m+1,m} & b_{m+1,m+1} & \cdots & b_{m+1,n} \\ \cdots & \cdots & \cdots & \cdots & \cdots & \cdots \\ b_{n,1} & \cdots & b_{n,m} & b_{n,m+1} & \cdots & b_{n,n} \end{array}\right] = \begin{bmatrix} \mathbf{B}_{11} & \mathbf{B}_{12} \\ \mathbf{B}_{21} & \mathbf{B}_{22} \end{bmatrix} \tag{8}$$

From the probit-based SUE assignment, both the mean value of link flow $\bar{\mathbf{v}}$ and the matrix of link flow variance/covariance $\mathbf{B}$ can be obtained. On the other hand, the actual link flow $\hat{\mathbf{v}}_\mathbf{d}$ is available on the links with detectors. However, a difference between $\hat{\mathbf{v}}_\mathbf{d}$ and $\bar{\mathbf{v}}_\mathbf{d}$ may exist due to various reasons such as inaccurate estimation of the OD matrix. The mean and variance of link flows on those links without detectors can be updated on the basis of the actual link flows obtained by the detectors. The mean value of link flow $\tilde{\mathbf{v}}_{\mathbf{e}}$ can then be estimated by Equation (5).

In the proposed TFS, the perceived travel time on the *k*th path connecting OD pair *rs* is assumed to follow a normal distribution of $N\left(c_k^{rs}, \omega c_k^{rs}\right)$, where c_k^{rs} is the actual path travel time and ω is the perception error coefficient, which will be calibrated in this chapter. The perception error coefficient is assumed to be known for application of the TFS. Note that different values taken by ω will result in different route choice patterns of drivers. The larger the parameter ω, the more diversified the route choice pattern.

On the other hand, the OD flows are not constant in real life due to stochastic variation in travel demand over time and/or the inconsistency of travel time with the time traffic counts obtained. It is assumed in the TFS that all the OD flows follow a normal distribution of

$$Q_{rs} \sim N\left(q_{rs}, \beta q_{rs}\right) \tag{9}$$

where β is the OD variation coefficient. Such OD fluctuation has also been considered in the road network reliability analysis by Asakura and Kashiwadani (1991).

Apart from assessing the travel time reliability, the TFS can also provide the estimated link flows for the whole network, link/path travel times, together with their variances and covariances. In addition, the prior OD matrix can be updated simultaneously. In order to apply the TFS model in practice, the perception error coefficient ω and the OD variation coefficient β have to be calibrated in advance.

CALIBRATION MEASURES

The selection of suitable measures for the model calibration plays an important role in the whole calibration process. The suitable measures vary from one problem to the next due to the different characteristics of the problem under study. For the calibration of a single parameter, the best measure(s) should be a monotonic function of the parameter that is to be calibrated so as to guarantee that the parameter and the value of the chosen measure have a one-to-one mapping. Otherwise, two or more different parameter values may lead to the same result of the chosen measure. Hence, the value of a parameter cannot be determined uniquely if the selected measure is not sufficient for the calibration.

A combined trip distribution and assignment (CDA) model for networks with multi-user classes was calibrated by Lam and Huang (1992). It was shown that three quantities, the mean trip cost, the OD trip entropy and the integral network cost, could be used individually for the model calibration. The OD trip entropy was found to be the best measure for the CDA model. The best measure that can be used for calibration of the TFS will be determined through a study of the characteristics of the following five measures.

The first measure is the integral network cost defined as:

$$c_n = \sum_a \int_0^{v_a} t_a(x)dx \tag{10}$$

In fact, this is the objective function for the deterministic user equilibrium (DUE) model and is equivalent to the third item in the objective function (1) of TFS. It is noted that the integral network cost is a decreasing function with respect to the parameter to be calibrated for the CDA model as reported by Lam and Huang (1992). However, there is no evidence or rigorous theoretical proof that c_n is a monotonic function of the perception error coefficient ω.

The second measure is the total trip cost of a network as shown below:

$$c_t = \sum_a v_a t(v_a) \tag{11}$$

It is the second item in the objective function (1) of TFS and is often used as the objective function for the traffic system optimization (SO) problem. It was found by Lam and Huang (1992) that the total trip cost will decrease when the dispersion parameter α of the CDA model increases. The increase of the dispersion parameter α in the logit-based CDA model is equivalent to 'reduction of perception error' that can be modeled by a decrease of the perception error coefficient ω in probit-based models. However, Maher and Hughes (1996) reported that in their probit model the reduction of the link-based dispersion coefficient does not always lead to a reduction in the total trip cost. In their example, the total trip cost increased while the dispersion coefficient decreased from 0.1 to 0. As defined before, ω is a path-based perception error coefficient and is not the same as the link-based dispersion coefficient employed by Maher and Hughes (1996). As the total trip cost may not be a monotonic function of ω in the probit-based model, tests should therefore be carried out to determine whether the total trip cost can be used as a suitable measure for the TFS calibration.

Entropy is the best measure suggested for calibration of the CDA model (Lam and Huang, 1992). Under user equilibrium (UE) conditions, no driver can find an alternative path with less travel time than the path he has chosen. Intuitively, the whole system is extremely stable under UE conditions and entropy should be minimised. Due to the perception error in the SUE condition, some drivers do not choose the shortest path and the system status varies from a stable one to some stochastic cases. If the perception error coefficient approaches infinity, the status of the system is totally unstable. This may correspond to a maximum entropy. It is expected that the entropy is a monotonic function with respect to the perception error coefficient. There are two types of entropy to be tested in this chapter, the link choice entropy and the path choice entropy.

The link choice entropy which is the third measure can be simply defined as

$$h_l = -\sum_a v_a \ln v_a \tag{12}$$

The path choice entropy can be calculated based on the path flow information. However, such information is unavailable in the TFS, as the TFS is basically aimed to obtain/store the link flow information. Akamatsu (1997) proposed a decomposition formula as shown below to calculate the path choice entropy based on the link flow information. The path choice entropy, which is the fourth measure, is decomposed from the most likely link flow patterns over the network as follows:

$$h_p = -\sum_a v_a \ln v_a + \sum_j \left(\sum_i v_{ij} \right) \ln \left(\sum_i v_{ij} \right) \tag{13}$$

where v_{ij} represents the flow on the link from node i to node j and $\sum_i v_{ij}$ is the total flows entering node j.

The above four measures are based on the mean values of link flows that are mainly affected by the perception error coefficient ω. On the other hand, the OD variation coefficient β will affect the variations of link flows but with little impact on the mean values of link flows due to the normal distribution given by Equation (9). Therefore the measures for calibrating β should be related to the variation of link flows. The network coefficient of variation (NCV), which was used by Asakura and Kashiwadani (1991) for calibration of link flow pattern, is considered as a measure for calibration of β in this chapter.

In the TFS, the traffic flow on link a follows a normal distribution of $N\left(v_a, \sigma_a^2\right)$. Then the NCV, which is the fifth measure, can be defined as

$$NCV = \sqrt{\sum_a \sigma_a^2 \Big/ \sum_a v_a} \tag{14}$$

The NCV provides an indication of link flow variation throughout the whole network.

The five measures proposed in this section will be examined, with an example used to investigate the best measures that can be applied for calibration of the TFS. It is easy to calibrate a single parameter of TFS, ω or β, by choosing a suitable measure and using a one-dimension search technique. However, the TFS calibration is a complicated problem since two parameters have to be calibrated simultaneously. It may be difficult to find a suitable measure that is monotonic with respect to both of the parameters. Therefore, we have to choose measures that are suitable for the two parameters separately. The least square formula can be used to integrate these measures to calibrate ω and β simultaneously. Under such circumstances, the traditional one-dimension search technique cannot be applied directly as it is difficult to find search directions for the two parameters respectively based on the least square values. Therefore a stochastic search technique is considered and a genetic algorithm (GA) is adopted for the calibration of TFS in this chapter.

GENETEC ALGORITHM

Genetic algorithms are search algorithms that are developed on the basis of natural selection and natural genetics. In the GA approach, it is firstly to code the parameters (or decision variables) into a string format and then to search the optimal solution among a stochastic population of the strings. One of the advantages of GA is to use the fitness function instead of the complicated knowledge required by traditional methods, such as continuity, derivative existence and monotonicity. A brief introduction to the GA operators used in this chapter is given below, before the description of the calibration procedures. For detailed information of GA, readers may refer to Goldberg (1989).

The first step when using GA is to code the perception error coefficient ω and the OD variation coefficient β as a binary finite-length string. Each coded string is called a chromosome and consists of a list of genes. In the binary coding employed in this chapter, each gene can either be 0 or 1. Each chromosome can be decoded and mapped to a certain value of ω and β. If the feasible solution of ω is assumed to fall in the range of $[\omega_{min}, \omega_{max}]$, the length of the chromosome for coding of ω is b, then the value of ω corresponding to a string of 1011 (the decimal integer is 11) can be decoded as

$$\omega = \omega_{min} + \frac{11}{2^b}(\omega_{max} - \omega_{min}) \tag{15}$$

The length of the chromosome will influence the accuracy of the solution as well as the convergence of the algorithm. A decoding equation similar to Equation (15) can be applied for decoding β accordingly. In the TFS calibration, the same length b is adopted for coding both ω and β.

In the operation of the GA, each chromosome in the parent generation is evaluated to obtain their fitness with respect to an objective function. With the fitness values of all the chromosomes in the parent generation, three operators (reproduction, crossover and mutation) are used to generate new chromosomes, which will then produce the child generation. Considering that the objective of the TFS calibration is to select ω and β that can best fit the OD matrix and the actual link counts together with their variations, the fitness can be defined as below.

$$f(\omega,\beta)=u-\left(M(\omega,\beta)-\hat{M}\right)^2 \tag{16}$$

In Equation (16), $M(\omega,\beta)$ is the value of the selected measure corresponding to ω and β. $\hat{M}$ is the value of the selected measure corresponding to the actual information or may be referred to as the target value of the selected measure, and u is a constant to ensure that the fitness is always positive. The values of ω and β corresponding to the maximum fitness value $f(\omega,\beta)$ are the optimum calibrated parameters.

Reproduction is an operation process through which chromosomes are copied into the mating pool with a probability proportional to their fitness. A hybrid model based on the combination of tournament selection and an elitist model is adopted in this study. In the parent generation, two chromosomes will be selected randomly. The one with the higher fitness will be copied into the mating pool. The reproduction process will be carried out repeatedly until the number of chromosomes in the mating pool reaches the predetermined population size z. If the chromosome in the parent generation with the highest fitness is not reproduced, then it will be copied into the mating pool.

The crossover operator provides search capability in the GA. The uniform crossover operator is adopted in this chapter. Two chromosomes in the mating pool will be randomly selected. A binary crossover mask with the same length as that of the chromosomes in the mating pool will be randomly generated according to a predetermined crossover probability P_c. Then the genes of the crossover mask will be scanned one by one. If a gene in the crossover mask is "1", the corresponding genes for the selected pair of chromosomes are exchanged; otherwise, they remain unchanged.

After the reproduction and crossover processes, the strings or the parameters in the child generation are improved. But these two processes may not lead to reliable results, since the optimal solution may be trapped into a local optimum. Mutation is a process used to overcome the above problem by the possibility of jumping out from a local optimum. The mutation process is done at the level of genes on the chromosomes obtained after the crossover. Each gene of a selected chromosome is allowed to mutate to the other possible value (i.e., in binary coding, if the gene is 0, it will change to 1) with a certain probability known as the mutation probability P_m.

CALIBRATION OF TFS

With the integration of GA approach with TFS, the proposed calibration procedure can be described in the following steps.

Step 1. Initialize. Set generation count g=1, population count n=1.
Step 2. Generate the initial population of chromosomes.
Step 3. Decode the n-th chromosome to get the perception error coefficient ω and the OD variation coefficient β.
Step 4. Running TFS with ω and β.
Step 5. Calculate the fitness of the present chromosome.
Step 6. If n equals the population size z, go to Step 7; otherwise, n=n+1, return to Step 3.
Step 7. If g equals the pre-determined maximum generation, go to Step 8; otherwise, generate the $g+1$th generation by reproduction, crossover and mutation operators, set g=g+1 and n=1, then return to Step 3.
Step 8. Find the chromosome with the highest fitness value, the decoded parameter is the result of the calibration.

The above calibration algorithm is illustrated in Figure 1 and will be tested using an example network in the next section.

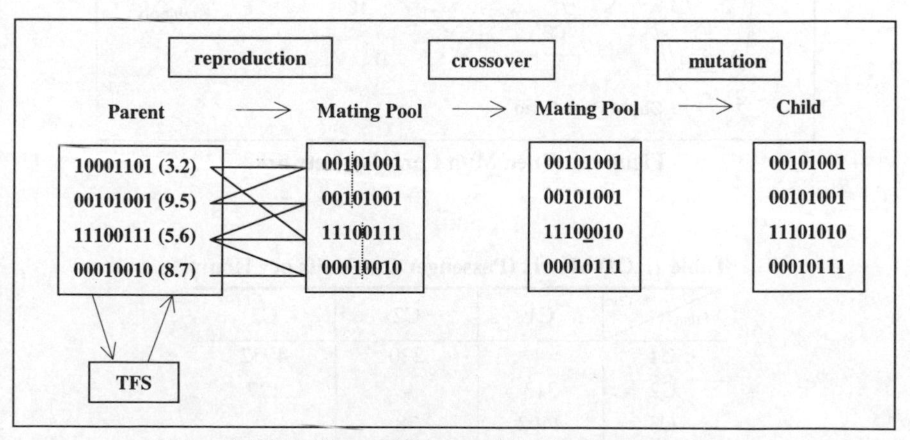

Figure 1. Flow Chart of GA-based Calibration Algorithm.

NUMERICAL EXAMPLE

The Tuen Mun corridor network in Hong Kong is used in this study for examining the effectiveness of the proposed GA-based calibration algorithm. As shown in Figure 2, the network consists of 10 links and 4 nodes, in which 3 nodes are connected to the three zone centroids. The Bureau of Public Road (BPR) function is adopted as the link travel time function:

$$t_a(v_a) = t_a(0) + k\left(\frac{v_a}{s_a}\right)^p \tag{17}$$

The OD matrix is shown in Table 1. The relevant link data and the observed traffic flows are given in Table 2. Note that the observed link flows are obtained by the TFS with the perception error coefficient $\omega = 0.3$ and the OD variation coefficient $\beta = 0.5$.

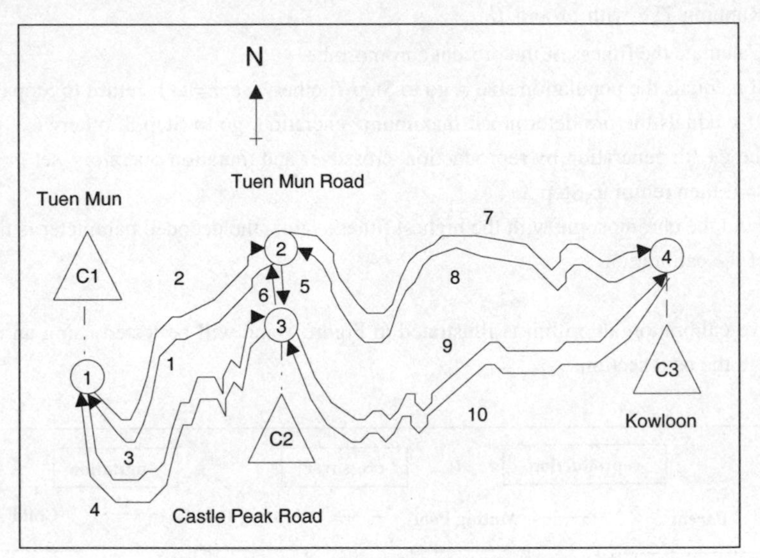

Figure 2. Tuen Mun Corridor Network.

Table 1. OD Matrix (Passenger Car Units per Hour).

From \ To	C1	C2	C3
C1	-	220	4952
C2	313	-	127
C3	4492	78	-

Table 2. The Link Data of the Network.

Link No.	$t_a(0)$ (hrs)	s_a (pcu/hr)	Parameters p	k	Observed link flow (pcu/hr)
1	0.0900	5175	3.5	0.1050	3360
2	0.0900	5175	3.5	0.1050	3687
3	0.1106	850	3.6	0.1408	1491
4	0.1106	850	3.6	0.1408	1451
5	0.0056	1150	3.6	0.0071	1476
6	0.0056	1150	3.6	0.0071	1569
7	0.0335	4800	3.6	0.0335	3743
8	0.0335	4800	3.6	0.0335	3323
9	0.0767	1000	3.6	0.1073	1368
10	0.0767	1000	3.6	0.1073	1279

Path Choice Entropy and NCV - the Best Measures for Calibration

The proposed calibration measures are tested with this example. Figure 3 shows the integral network cost with different ω in the example network. The integral network cost tends to increase with increasing ω when β remains unchanged. So it can be used as an alternative measure.

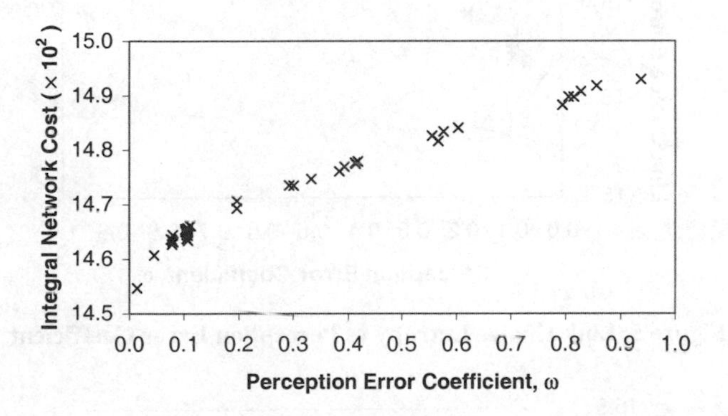

Figure 3. Integral Network Cost vs Perception Error Coefficient.

The total trip cost of the example network with various ω is plotted in Figure 4. It is obvious that the total trip cost is not a monotonic function of ω. When ω increases from 0 to 0.3, the total trip cost decreases, This pattern is similar to the result reported by Maher and Hughes (1996). On the other hand, the total trip cost increases when the value of ω increases from 0.3 to 1.0. As a result, two different ω may have the same total trip cost. Therefore the total trip cost should not be used for the TFS calibration as it is preferred to have a monotonic relationship between the measure and the parameter.

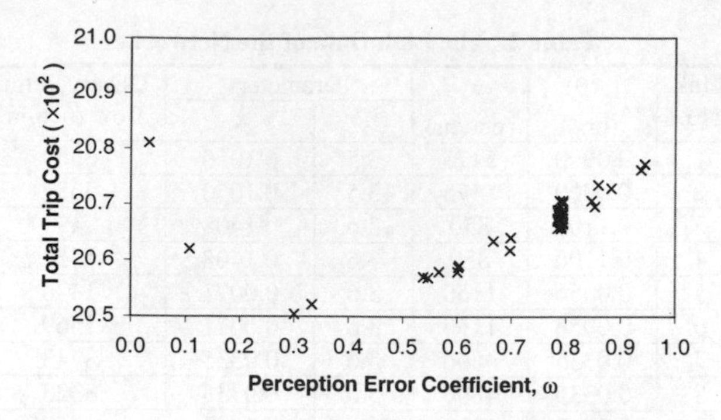

Figure 4. Total Trip Cost vs Perception Error Coefficient.

Both the link choice and path choice entropy have been tested. The results are presented in Figures 5 and 6 respectively. It can be seen that both entropy indices show monotonic tendency with respect to the perception error coefficient ω. And the curve of the path choice entropy seems to be more stable than that of the link choice entropy.

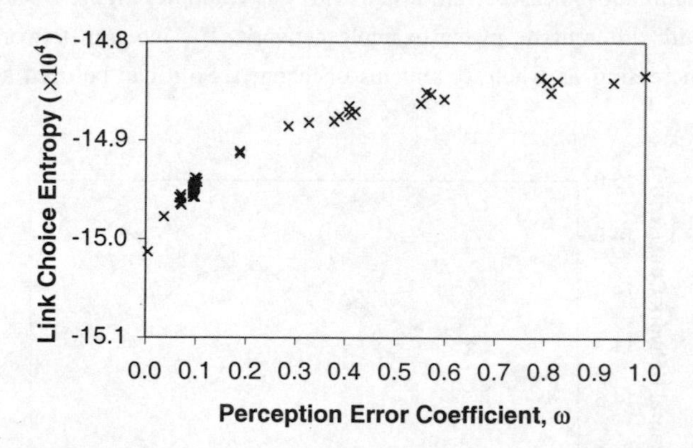

Figure 5. Link Choice Entropy vs Perception Error Coefficient.

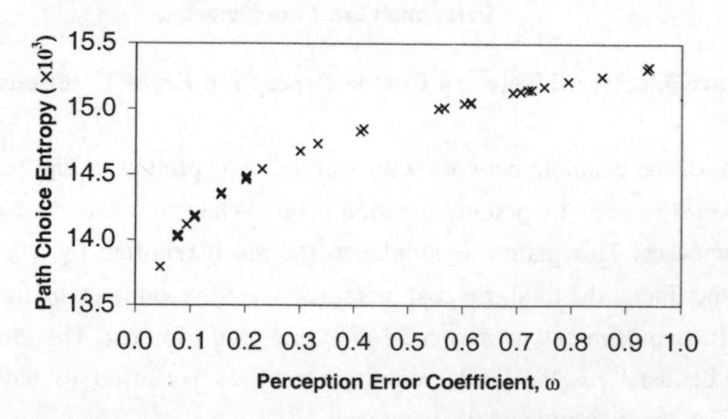

Figure 6. Path Choice Entropy vs Perception Error Coefficient.

The NCV values corresponding to different values of β are plotted in Figure 7. It is clear that the NCV is a monotonic function with respect to β, and therefore is suitable for the calibration of β in the TFS.

Based on the above results, it can be found that the path choice entropy and NCV are the best measures for the calibration of TFS in the example network provided that all the link flow information can be available. However, it may be difficult and expensive to collect the complete traffic flow data. In practice, partial flow information is usually available. In view of this, the combination of integral network cost and NCV would be the better alternative if the objective path choice entropy can not be obtained particularly when the complete link flow information is not available.

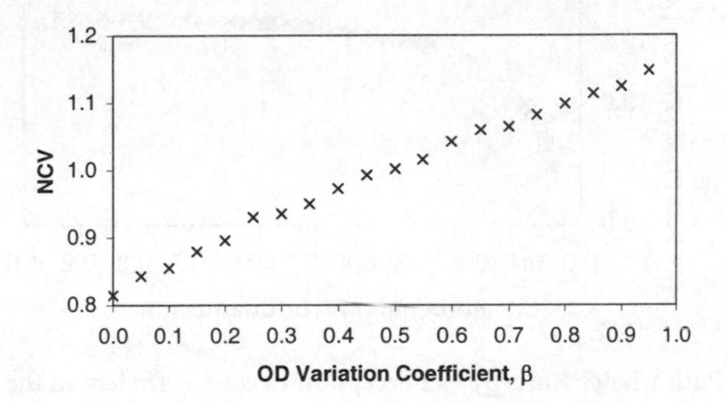

Figure 7. NCV vs OD Variation Coefficient.

Calibration Results

The discussion on the calibration measures in the previous section is based on the assumption that one parameter changes while the other one remains unchanged. However, in the TFS calibration, two parameters ω and β will be calibrated simultaneously.

By integration of the path choice entropy and NCV as the calibration measures, the fitness function (16) can be written as

$$f = u - \eta_1 \left(\frac{h_p(\omega,\beta) - \hat{h}_p}{\hat{h}_p} \right)^2 - \eta_2 \left(\frac{NCV(\omega,\beta) - NCV_o}{NCV_o} \right)^2 - \eta_3 \sum_{rs} \left(\frac{q_{rs} - \hat{q}_{rs}}{\hat{q}_{rs}} \right)^2 \tag{18}$$

where η_1, η_2 and η_3 are the coefficients to scale the corresponding least square terms within the range of $[0,1]$. These three coefficients can be determined by the following equations.

$$\eta_1 = \left(\frac{h_p^{\min}}{h_p^{\max} - h_p^{\min}}\right)^2, \ \eta_2 = \left(\frac{NCV^{\min}}{NCV^{\max} - NCV^{\min}}\right)^2 \text{ and } \eta_3 = \left(\frac{q_{rs}^{\min}}{q_{rs}^{\max} - q_{rs}^{\min}}\right)^2 \qquad (19)$$

In order to determine the values of these coefficients, pilot tests on TFS are undertaken to find the approximate lower and upper limits of the path choice entropy and NCV. The path choice entropy with respect to ω is plotted in Figure 8. It can be seen that the monotonicity is not violated although the OD variation coefficient β is not fixed. And the maximum path choice entropy is about 20000 while the minimum is about 18000. Therefore the value of η_1 is set to 100.

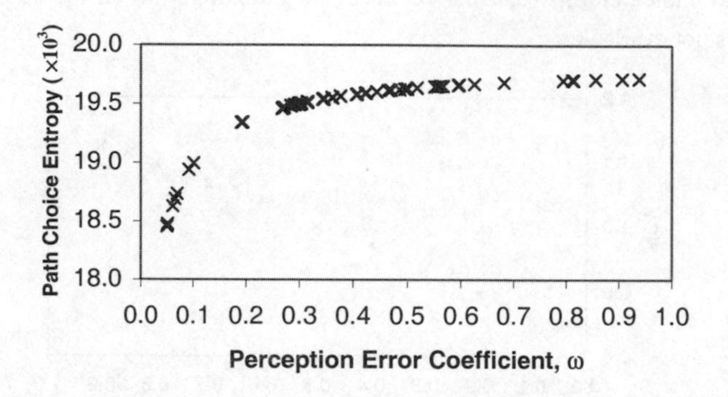

Figure 8. Path Choice Entropy vs Perception Error Coefficient in the Pilot Tests.

Figure 9 shows the relationship between NCV and β. It is obvious that the NCV is not a strictly monotonic function of β in the calibration process due to the influence of various ω. According to Equation (19), η_2 is initially set to be 0.25. By combining constraint (4) and Equation (19), $\eta_3 = \left(\frac{1-\delta}{2\delta}\right)^2$. Assuming $\delta = 0.5$ in the numerical example, $\eta_3 = 0.25$.

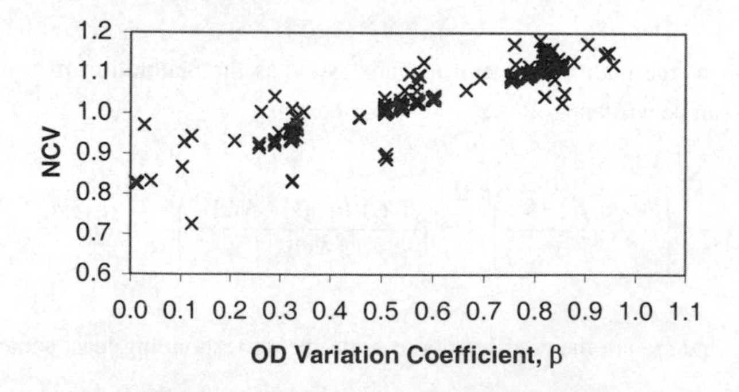

Figure 9. NCV vs OD Variation Coefficient in the Pilot Tests.

Based on the initial values of η_1, η_2 and η_3 as shown above, it has been found that $u = 0.1$ can guarantee that the fitness value is always positive, ranging from 0 to 0.1. A multiplier of 1000 is used to amplify the fitness value to the range of $[0,100]$. Therefore the values for these coefficients in the TFS calibration are $u = 100$, $\eta_1 = 10^5$ and $\eta_2 = \eta_3 = 250$. It should be noted that the determination of these parameters is based on the assumption that the information on traffic count and NCV is accurate. However, it is generally believed that the accuracy of traffic counts is greater than that of NCV. Hence, a greater value should be taken for η_1.

Before carrying out the calibration, some parameters in the GA should be determined in advance. Such parameters include population size, length of chromosome, maximum number of generations, crossover probability and mutation probability. Several tests have been carried out with different combinations of population size, *z*, and chromosome length for coding each parameter, *b*. The resultant maximum fitness values are presented in Figure 10.

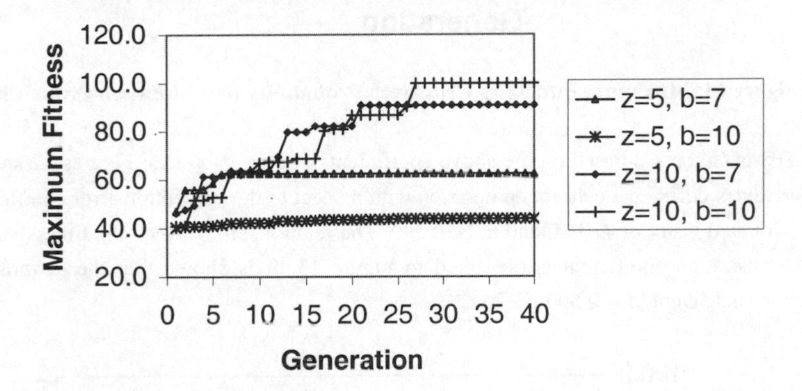

Figure 10. Maximum Fitness vs Population Size, Generation, Length of Chromosome.

In Figure 10, it can be seen that the test result of the combination of a population size of 10 with a length of 10 bits for each parameter leads to the largest fitness value of all three combinations. Therefore, the population size is set to be 10 and the maximum generation is set to be 30. Each parameter will be coded into a 10-bit length binary string and the chromosome will then have a total length of 20 bits.

Similarly, various combinations of crossover probability P_c and mutation probability P_m have been tested to find the suitable values of the captioned parameters for the example. The results of these tests are shown in Figure 11.

In Figure 11, it is obvious that the combination with the crossover probability of 0.5 and the mutation probability of 0.02 shows better result than the other four combinations. Therefore the crossover probability is set to be 0.5 and the mutation probability is set to be 0.02. The hybrid reproduction operator described in the above section is adopted. The search space is $\omega \in [0,1]$ and $\beta \in [0,1]$.

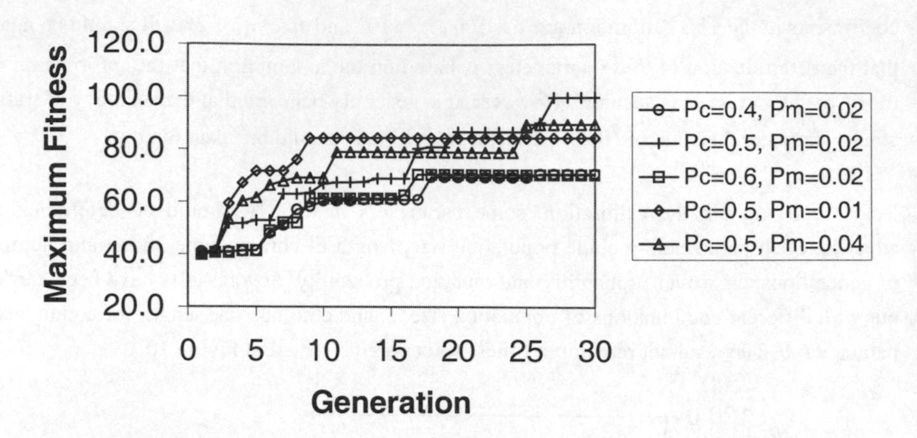

Figure 11. Maximum Fitness vs Crossover Probability and Mutation Probability.

The TFS is calibrated based on the above coefficients and probabilities. Figures 12 shows the fitness values during the calibration iteration with respect to the perception error coefficient ω. The calibrated result of ω is found to be 0.299. The relationship between the fitness value and the OD variation coefficient is presented in Figure 13. It is shown that the calibrated OD variation coefficient $\beta = 0.509$.

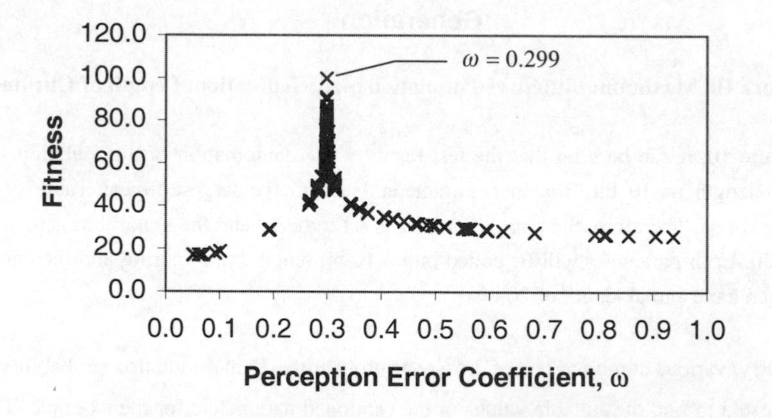

Figure 12. Fitness vs Perception Error Coefficient in the TFS Calibration.

The calibrated results of the two parameters are very close to their actual values $\alpha = 0.3$ and $\beta = 0.5$ respectively. The maximum relative calibration error is 1.8% only. So it can be concluded that the GA-based calibration algorithm can be used for the TFS calibration as satisfactory results were found in the numerical example.

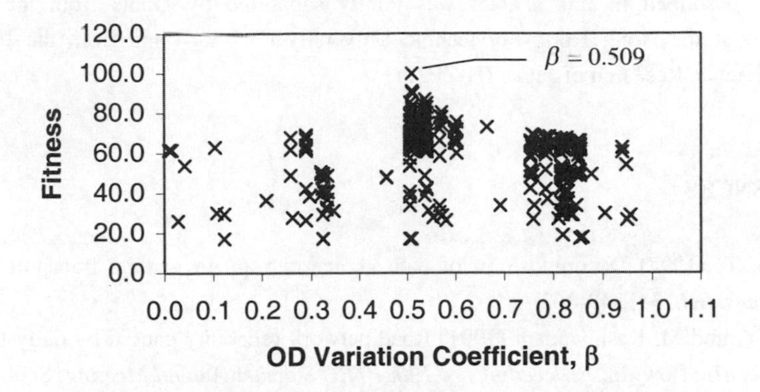

Figure 13. Fitness vs OD Variation Coefficient in the TFS Calibration.

CONCLUSIONS AND RECOMMENDATIONS

A TFS has recently been proposed for the assessment of network reliability in terms of travel time reliability. However, the perception error coefficient (ω) and the OD variation coefficient (β) need to be calibrated before the application of the model. In view of the complexity of the constrained probit-type SUE model in the TFS, a new algorithm based on GA has been developed in this chapter. The proposed calibration algorithm has been tested on an example network. The calibration results are promising and the performance of the proposed calibration measures is illustrated. It is found that the path choice entropy and NCV are the best measures for the TFS calibration. The GA parameters, such as population size, maximum generations, crossover probability and mutation probability are optimized before the TFS calibration. Therefore, it can be concluded that the proposed GA-based method is suitable for the TFS calibration.

Further tests will be carried out with large-scale networks to verify the effectiveness of the proposed calibration algorithm. Extensions of TFS to multiple user classes and time-dependent dimensions are being carried out. In order to assess the network reliability improvements due to the Advanced Traveller Information System (ATIS), multiple user classes should be taken into account in the reliability analysis so as to differentiate the drivers with different levels of traffic information. Further effort will therefore be required to calibrate the TFS with multiple user classes in time-dependent network in order to make the TFS more applicable for evaluation of alternative ATIS options.

ACKNOWLEDGEMENTS

The work described in this chapter was jointly supported by grants from the Research Committee of the Hong Kong Polytechnic University (G-V586) and from the Institute of Systems Science Research of Japan (H-ZF21).

REFERENCES

Akamatsu, T. (1997) Decomposition of path choice entropy in general transport networks. *Trans. Sci.*, **31**, 349-362.

Asakura, Y. and M. Kashiwadani (1991) Road network reliability caused by daily fluctuation of traffic flow. In: *Proceedings of 19th PTRC Summer Annual Meeting*, Seminar G, pp. 73-84. PTRC Education and Research Services Ltd., London.

Bell, M.G.H. and Y. Iida (1997) *Transportation Network Analysis*, Wiley, Chichester.

Bolduc, D. (1999) A practical technique to estimate multinomial probit models in transportation. *Trans. Res.-B.*, **33**, 63-79.

Cascetta, E., A. Nuzzolo, F. Fusso and A. Vitetta (1996) A modified logit route choice model overcoming path overlapping problems, specification and some calibration results for interurban networks. In: *Proceedings of the 13th International Symposium on Transportation and Traffic Theory* (J.B. Lesort, ed.), pp. 697-711. Pergamon, Oxford.

Daganzo, C.F. (1979) *Multinomial Probit: the theory and its application to demand forecasting*, Academic Press, New York.

Ghareib, A.H. (1996) Evaluation of logit and probit models in mode-choice situation. *J. Trans. Engg.*, **122(4)**, 282-290.

Goldberg, D.E. (1989) *Genetic Algorithms in Search, Optimization & Machine Learning*. Addison-Wesley Publishing Company, Massachusetts.

Iida, Y. (1999) Basic concepts and future directions of road network reliability analysis. *J. Adv. Trans.*, **33(2)**, 125-134.

Lam, W.H.K. and H.J. Huang (1992) Calibration of the combined trip distribution and assignment model for multiple user classes. *Trans. Res.-B*, **26**, 289-305.

Lam, W.H.K. and G. Xu (1999) A traffic flow simulator for network reliability assessment. *J. Adv. Trans.*, **33(2)**, 159-182.

Maher, M.J. (1992) SAM: a stochastic assignment model. In: *Mathematics in Transport Planning and Control* (J.D. Griffiths, ed.), pp. 121-131. Clarendon Press, Oxford.

Maher, M.J. and P.C. Hughes (1996) Estimation of the potential benefits from an ATT system using a multiple user class stochastic user equilibrium assignment model. In: *Proceedings of the Fourth International Conference on Applications of Advanced Technologies in Transportation Engineering* (Y.J. Stephanedes & F. Filippi, ed.), pp. 700-704. ASCE, New York.

Sparmann, J.M., C.F. Daganzo and M. Soheily (1983) Linear probit models: statistical properties and improved estimation methods. *Trans. Res.-B.*, **17(1)**, 67-86.

Wong, C.K., S.C. Wong and C.O. Tong (1998) A new methodology for calibrating the Lowry model. *J. of Urban Planning and Development*, **124(2)**, 72-91.

Yai, T., S. Iwakura and S. Morichi (1997) Multinomial probit with structured covariance for route choice behavior. *Trans. Res.-B.*, **31(3)**, 195-207.

APPENDIX: NOTATIONS

$\mathbf{B}$: variance/covariance matrix of link flows.

c_k^{rs} : actual travel time on path *k* between OD pair *rs*.

k : parameter in BPR function.

p : parameter in BPR function.

p_a^{rs} : link choice proportion for link *a* of OD pair *rs* (i.e. proportion of flow from *r* to *s* using link *a*). According to the links equipped with detector or not, the link choice proportion can be divided into two subsets of p_d^{rs} and p_e^{rs}.

q_{rs} : mean OD flow.

$\hat{q}_{rs}$: prior OD flow.

S_{rs} : expected minimum travel time between OD pair *rs*.

s_a : capacity of link *a*.

$t_a(0)$:free flow travel time on link *a*.

$t_a(v_a)$:travel time on link *a* with traffic flow of v_a.

v_a : traffic flow on link *a*.

$\bar{v}_d$: mean flow on link with detector.

$\hat{v}_d$: detected link flow.

$\bar{v}_e$: mean flow on link without detector.

$\tilde{v}_e$: estimated flow on link without detector based on the detected data.

β : coefficient of variance for OD variation.

ω : coefficient of variance for perception error of path travel time.

δ : tolerance parameter of OD flows.

CHAPTER 11

A CHANCE CONSTRAINED NETWORK CAPACITY MODEL

Hong K. Lo[1] *and Y. K. Tung*[2]

Department of Civil Engineering, Hong Kong University of Science and Technology
Clear Water Bay, Hong Kong, China.

1. INTRODUCTION

The study of the reliability of a road network recently has become an important research topic. The reason is simple. Disruptions of the road network have a significant impact on the economy of a nation or region. The recent Kobe earthquake in Japan and the Loma Prieta earthquake in the United States are such examples that have raised concern on this issue. Particularly, the roadways in the Kobe region were so severely disrupted that mobility within the area was almost completely disabled (Wakabayashi, 1996). This demonstrates the vulnerability of transportation systems to natural disasters. On a day-to-day basis, disruptions on a minor scale occur in the form of traffic incidents, which also degrade the performance of a road network. It is therefore an important research topic to understand and, if possible, quantify such impacts. Such an understanding may help to improve network design.

In the past, reliability analysis has received very little attention in the study of road networks in spite of its importance. The few existing reliability studies of road networks are mainly limited to two aspects: connectivity and travel time reliability (Bell and Iida, 1997). Connectivity reliability is concerned with the probability that the network nodes remain connected. A special case of the connectivity reliability is the terminal reliability, which concerns the existence of a route between a specific origin-destination (O-D) pair (Iida and Wakabayashi, 1989). For each node pair, the network is considered successful if at least one route is operational. Capacity constraints on the links are not accounted for when finding the connectivity reliability.

Another measure of network reliability is travel time reliability or the coefficient of variation of travel time (Asakura and Kashiwadani, 1991). This is concerned with the probability that a trip between a given O-D pair can be made successfully within a specified interval of time. This measure is useful to evaluate network performance under normal daily flow variations. Bell et al. (1998) proposed a sensitivity analysis based procedure to estimate the variance of travel

[1] cehklo@ust.hk
[2] cetung@ust.hk

time arising from daily demand fluctuations. Asakura (1996) extended the travel time reliability to consider capacity degradation due to deteriorated roads. He defined travel time reliability as a function of the ratio of travel times under the degraded and non-degraded states.

Recently, Du and Nicholson (1997) proposed a theoretical framework for analysis of degradable transportation systems. A conventional integrated network equilibrium model with variable demand is used to describe flow on a degradable transport network whose components (roadway capacities) are subject to a range of levels of degradation. System surplus is used as a performance measure to assess the socio-economic impacts of system degradation. The reliability of interest is the probability that the reduction in flow of the system is less than a certain threshold, as a result of capacity degradation of the network.

The existing reliability measures are useful to assess different factors related to the performance of the transportation network. However, none of the above measures addresses the issue of adequacy of network capacity to accommodate demand. That is, whether the available network capacity relative to the required demand is sufficient. This measure provides important information for efficient flow control, capacity expansion, and other relevant works, thus enhancing the reliability of a road network. It may also have the potential of providing a tool to design road networks that are resistant to traffic disturbances.

Capacity reliability of a network can be considered in two ways. Recently, Chen et al. (1999a, 1999b) defined capacity reliability as the probability that the network can accommodate a certain traffic demand at a required service level. They developed an assessment methodology, based on sensitivity analysis and Monte Carlo methods, to evaluate the performance of a degradable road network. Though useful, this methodology requires a lot of computational power.

Capacity reliability of a network can also be considered in a simpler way, defined in this study as the maximum flow that the network can carry subject to the link capacity and/or travel time reliability constraints. Traffic engineers know a lot more about the performance of links, in terms of capacity and travel time reliabilities. This is because most of the traffic statistics and accident records are collected at the link level. It is, therefore, natural to specify the required performances at the link level and to evaluate their eventual impact on the network. For example, one may specify for each link that its traffic flow should not exceed its capacity for more than 10% of the time. Since the capacity of a link is not fixed but subject to stochastic degradation, the above performance specification is stochastic or a chance constraint. The question is then what is the maximum flow a network can carry such that all the links fulfill this performance specification. The crux of this analysis relies on an approach referred to as chance-constrained programming problems (Charnes and Cooper, 1963). This approach has been applied extensively to the analysis of environmental planning, water system design, and utility distribution (examples, Guldmann, 1983; Fuessle et al., 1987; Mays and Tung, 1992; Jacobs, et al. 1997).

This chapter develops a formulation to evaluate network capacity that is based on two types of link reliabilities. One is link capacity reliability. The other is link travel time reliability. We will discuss both types of reliabilities and show how they can be converted to a deterministic equivalent. The result is that one only needs to solve the equivalent deterministic mathematical program once without the need for repeated Monte Carlo simulations, which is often computationally intensive. The details are discussed in Section 2.

The rest of this chapter is as follows. Section 3 is the numerical study. We apply the methodology to the Sioux Falls network and discuss the numerical results. Finally, Section 4 contains some concluding remarks.

2. NETWORK CAPACITY WITH CHANCE CONSTRAINTS

The objective of this formulation is to determine the maximum attainable flow that a network can carry subject to the link reliability constraints. Consider a transportation network modeled by a directed graph $G(N,A)$ where N is a set of nodes and A is a set of links. W is a subset of N, where travel demands either enter or leave the network. The links on the strongly connected graph are roadways that make up the transportation network. Each link $a \in A$[3] has a certain capacity (C_a) and the maximum capacity of the network is determined by the value of an output parameter (μ), which can be computed from the capacities of all the roadways:

$$\mu = g(C_1, C_2, \cdots, C_a) \tag{1}$$

For networks containing many origins and destinations, the function $g(\cdot)$ may not exist analytically but can be determined by an optimization procedure. Specifically, the maximum flow problem, which is to find a feasible flow that leads to maximum flow capacity, can be formulated as a linear program (Ahuja et al., 1993). This method has been used in communication networks (Aggarwal, 1985; Chan el al., 1997), water distribution systems (Li et al., 1993), electric power systems (Billington and Li, 1994), and others to determine the maximum flow capacity of the network. We note that the capacity of a road network depends not only on the link capacities but also on route choice (Yang and Bell, 1998a). However, as a first effort in this modeling approach, we leave out route choice considerations. Such omission simplifies the formulations considerably. Nevertheless, it can be achieved within the proposed modeling framework, which we leave as an extension to this study. Thus, this formulation provides an upper bound solution to the analysis of the capacity of a road network.

Many aspects of the operations of a transportation network inevitably involve uncertainties. In particular, they could arise from the following sources: (i) variations in link capacities; (ii) variations in travel demands; (iii) imperfection of the route choice models; (iv) uncertainties in

[3] To more clearly specify the start and end nodes i, $j \in N$, respectively, of link a, in the later-on sections, we also use the notation (i,j) or simply ij to represent link a.

the parameters and forms of the impedance function. This study focuses on the impact on the maximum network capacity due to stochastic variations of the link capacities. Travel demands are implicitly accounted for while determining the maximum capacity of the road network. The uncertainties due to route choice models and impedance's parameters may also play a role in the estimation process. These issues could be extensions of this study.

There are many reasons why link capacities are subject to variations. In peak hours, congestion generally happens in urban areas due to large traffic demands that overload the system. The congestion could repeat almost daily and is thus referred to as a recurrent disturbance. Another kind of disturbance is induced by events such as earthquakes, floods, adverse weather, landslides or traffic accidents, which could cause temporary or prolonged disability on parts of the network such that the capacity would be substantially reduced. These events are referred to as non-recurrent disturbances. In any case, the link capacity should not be considered as a constant per se but should follow a probability distribution. Depending on the specificity of the link, the shape and functional form of the distribution could be different. Some possible examples are provided in Figure 1. In fact, these distributions could be estimated based on past accident records or other traffic statistics.

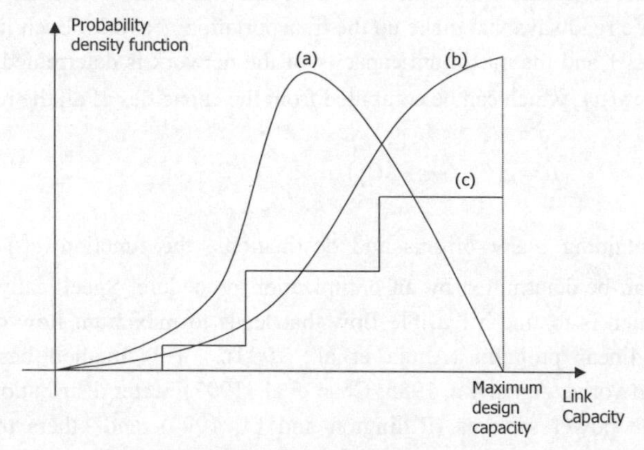

Figure 1 Possible probability density functions of link capacity.

2.1 Link Capacity and Travel Time Reliability Constraints

This study defines two types of link reliabilities. The first is referred to as capacity exceedance probability, defined as the probability that traffic flow on a link exceeds its capacity. Since link capacity is a random variable, one cannot state for sure that its capacity shall never be exceeded. Instead, we require the probability of such occurrence to be lower than a required or satisfactory level. This level could be defined by the traffic management agency, for example. Mathematically, it can be written as:

$$P\{x_a \geq C_a\} \leq \alpha_a, \tag{2}$$

where the subscript a refers to a particular link. The variables x_a, C_a, respectively, are the traffic flow and capacity of link a. The parameter α_a, referred to as the capacity exceedance probability for link a, defines the acceptable level of probability. Obviously, it could be different for different links. People generally tolerate higher congestion levels for links situated in business districts than links in residential areas near their homes for example.

In (2), C_a is a random variable specified by a particular probability density function (PDF) while x_a is the decision variable that is indirectly being maximized through the multiplier μ. The left-hand side (LHS) of (2) can be considered as a cumulative distribution function (CDF) of C_a, written as:

$$F_{C_a}(x_a) = P\{C_a \leq x_a\} \tag{3}$$

Using (3), (2) can be rewritten as:

$$F_{C_a}(x_a) \leq \alpha_a. \tag{4}$$

Since CDF are monotonic one-to-one functions, one can take the inverse of (4) and write:

$$x_a \leq F_{C_a}^{-1}(\alpha_a) \tag{5}$$

By specifying the CDF of the link capacity C_a and the acceptable capacity exceedance probability α_a, (5) becomes a deterministic constraint. Moreover, it is linear which makes it easy to be incorporated in an optimization problem.

The second type of reliability is referred to as the travel time exceedance probability, which takes a similar form as the link capacity reliability. It is defined as the probability that the link travel time exceeds a certain acceptable level. Mathematically, it is:

$$P\{t_a(x_a) \geq \phi_a \cdot t_a^f\} \leq \beta_a \tag{6}$$

where $t_a(x_a)$, t_a^f, respectively, are the travel time and free-flow travel time on link a. The parameters ϕ_a, β_a define the performance criteria. The former parameter is the congestion factor, which is basically greater than one since it defines the acceptable congestion above the free-flow travel. The latter parameter states the acceptable travel time exceedance probability. Using the BPR function as the link impedance function, (6) can be further simplified. Specifically, let

$$t_a(x_a) = t_a^f \left[1 + \tau \left(\frac{x_a}{C_a} \right)^{\rho} \right]. \tag{7}$$

where τ, ρ are parameters of the BPR function. Substituting (7) into (6) and simplifying, we obtain:

$$P\left\{ 1 + \tau \left(\frac{x_a}{C_a} \right)^{\rho} \geq \phi_a \right\} \leq \beta_a . \tag{8}$$

Rearranging, the expression becomes:

$$P\left\{ C_a \leq \frac{x_a}{\left(\dfrac{\phi_a - 1}{\tau} \right)^{1/\rho}} \right\} \leq \beta_a \tag{9}$$

Following the same logic as in the case of link capacity reliability, (9) can be stated as:

$$F_{C_a}\left(\frac{x_a}{\left(\dfrac{\phi_a - 1}{\tau} \right)^{1/\rho}} \right) \leq \beta_a \quad \text{or}$$
$$x_a \leq \left(\frac{\phi_a - 1}{\tau} \right)^{1/\rho} F_{C_a}^{-1}(\beta_a) \tag{10}$$

Even though (10) appears as a more complicated expression, it is still linear in x_a and is deterministic. With the parameters $\tau, \rho, \phi_a, \beta_a$, and the CDF of C_a fully specified, it is no more complicated than (5). In fact, since the CDF does not contain the decision variable x_a in these expressions, this approach can work with any distributions without any numerical or computational difficulty. These two reliability constraints are not mutually exclusive. If desired, they can be defined and satisfied simultaneously in a problem.

2.2 Network Capacity Model with Chance Constraints

The network capacity model's objective is to maximize the demand multiplier μ, subject to the nodal conservation constraints and the above reliability constraints. This concept of finding the maximum multiplier is referred to as the reserve capacity of a network (Wong and Yang, 1997). The program can be stated as:

$$\text{Max } \mu \tag{11}$$
$$\text{s.t.}$$
$$\sum_{\{j:(i,j)\in A\}} x_{ij} - \sum_{\{j:(j,i)\in A\}} x_{ji} = \mu \cdot [b(i)], \quad \forall i \in N \tag{12}$$

$$x_{ij} \le F_{C_{ij}}^{-1}(\alpha_{ij}), \quad \forall (i,j) \in A$$
$$x_{ij} \le \left(\frac{\phi_{ij} - 1}{\tau}\right)^{1/\rho} F_{C_{ij}}^{-1}(\beta_{ij}), \quad \forall (i,j) \in A \tag{13}$$

$$x_{ij} \ge 0, \quad \forall (i,j) \in A \tag{14}$$

In the notation here, link a is represented by its start and end nodes, i and j. That is, $a = (i,j) \in A$ where $i, j \in N$. The terms $\sum_{\{j:(i,j)\in A\}} x_{ij}$ and $\sum_{\{j:(j,i)\in A\}} x_{ji}$, respectively, represent the sum of flow out of and into node i. The variable $b(i)$ represents the original traffic that enters or leaves at node i. A positive (negative) value represents entering (leaving) traffic at node i. Constraint (12) defines the nodal conservation condition. The variable μ is the multiplier that is to be maximized while fulfilling either or both of the link reliability constraints (13). It should be observed that (11)-(14) is a linear program (LP) that can be solved readily. Physically, this LP finds the maximum demand a network can carry subject to the reliability constraints.

In solving the linear program (11)-(14), the dual variables or shadow prices associated with the constraints (13) will indicate the binding or sensitive links. To increase the network capacity, one can either increase the capacities of these sensitive links, change their capacity distributions (by reducing the incident occurrence rate for instance), or specify a higher tolerance (i.e. admitting a higher probability of the link's capacity or travel time level being exceeded). This information is valuable in pinpointing the congestion black spots for improvement considerations.

3. NUMERICAL STUDIES

To demonstrate this formulation, we apply it to the Sioux Falls network, as shown in Figure 2, with the full set of travel demands as described in LeBlanc et al. (1975). In this study, the capacity of each link is assumed to follow a beta distribution for its flexibility in defining various PDF shapes, as shown in Figure 3. In particular, by defining the upper and lower bounds of the random variable and two scaling parameters, commonly referred to as q and r parameters, beta distribution can fit various PDF shapes, ranging from uniform distribution to something that approximates a normal or gamma distribution. Mathematically, the PDF of beta distribution can be written as (Ang and Tang, 1990):

$$f_X(x)=\frac{1}{B(q,r)}\frac{(x-a)^{q-1}(b-x)^{r-1}}{(b-a)^{q+r-1}},\quad a\le x\le b \tag{15}$$

where X is the random variable following the beta distribution; a and b are, respectively, the lower and upper bound of the distribution; q and r are parameters to fit the PDF shape; $B(q,r)$ is the beta function defined as $B(q,r)=\int_0^1 x^{q-1}(1-x)^{r-1}dx$.

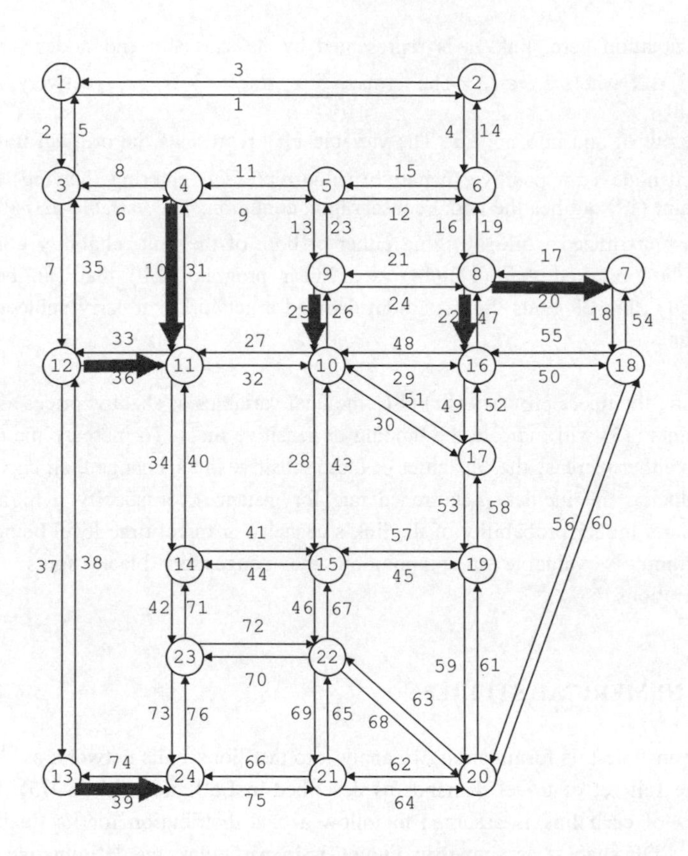

Figure 2 The Sioux Falls network.

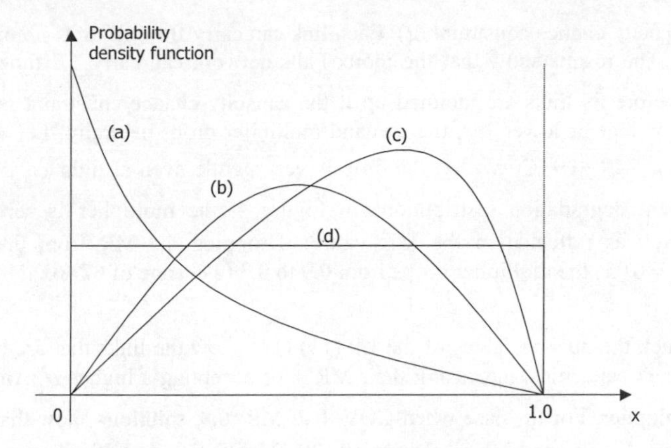

Figure 3 Various shapes of a standard beta PDF.

We set the upper bound to be each link's nominal capacity, CAP_{ij}, and the lower bound to be one quarter of the link's nominal capacity, $CAP_{ij}/4$. In other words, the capacity of each link varies randomly between its nominal capacity and one-quarter its nominal capacity while following the beta distribution. Of course, in reality, one can model different upper and lower bounds for each link. Moreover, to simplify exposition, instead of defining the q and r parameters, which are harder to interpret physically, we specify the mean and coefficient of variation (COV) of the capacity of each link and determine the corresponding beta distribution that produces this result. We emphasize that the beta distribution can provide a close fit to what the actual data look like. The values chosen here are simply for illustration purposes.

Two sets of results are presented here. The first set evaluates the performance of the network when only link capacity reliability is considered. The second set examines the impact of imposing the constraints of link travel time reliability.

Figure 4 shows the change in the demand multiplier μ with the capacity exceedance probability. COV refers to the coefficient of variation of the link capacity and MR refers to the ratio of the mean capacity of a link to its nominal capacity. A higher COV corresponds to a wider spread of the distribution. MR is always between zero and one. An MR close to one refers to the case where each link performs at its nominal capacity for most of the time. In reality, each link may have a different COV and MR. For simplicity, each link in this study follows the same COV and MR in each numerical study. The results for four different capacity degradation distributions are shown in Figure 4.

The acceptable capacity exceedance probability α_{ij} is specified by the user or the traffic management agency, which is set to be identical for each link (i, j) in this study but could be set differently for each link. When α_{ij} is close to 1, essentially, this is equivalent to relaxing or ignoring the capacity chance constraint (2). Each link can carry flow up to its nominal capacity, i.e. $x_{ij} \le CAP_{ij}$. The results show that the Sioux Falls network can carry 1.2 times its original demand level before its links are jammed up if the capacity chance constraint is ignored. As one specifies a tighter or lower α_{ij}, the demand multiplier drops naturally. For mild capacity degradations, e.g. MR=0.9, COV=0.1, the drop is very gentle even at tight α_{ij}'s. Comparing the four different degradation distributions in Figure 4, the multiplier is sensitive to the degradation level, as reflected in the MR values. Dropping the MR from 0.9 to 0.5, for example, at $\alpha_{ij} = 0.05$, the multiplier drops from 0.9 to 0.3 (a change of 67%).

As a side-product, the dual variables of the LP (11)-(14) show the links that are binding. That is, improving their capacities, increasing their MR's, or accepting a higher α_{ij} would increase the demand multiplier. For the case when COV=0.2, MR=0.8, solutions show that six links in the Sioux Falls network are binding: Links 10, 20, 22, 25, 36, and 39, shown as thickened arrows in Figure 2. They are candidates for improvement considerations. Figure 5 shows the increase in the multiplier as each and all of these links are improved by doubling their capacities, with the rest of the network unchanged. The results show that by improving all six links (out of a network of 76 links), the multiplier is increased by around 20% for all levels of α_{ij}. In particular, Links 10, 22, 36, and 39 can produce the biggest improvements with the existing travel demand pattern.

The next set of results shows the effect of imposing the travel time reliability on the demand multiplier. Figure 6 shows the contours of the multipliers for two sets of capacity degradation distributions, which indirectly affect the resultant link travel times. The top chart shows the case when COV=0.2, MR=0.8; while the bottom chart shows the case when COV =0.4, MR=0.5. In both charts, the horizontal axis refers to the travel time exceedance probability β_{ij} while the vertical axis is the congestion factor ϕ_{ij}.

With a small congestion factor ϕ_{ij} (say, around 1.5) and a tight travel time exceedance probability β_{ij}, (i.e., the region to the bottom-left region of the charts), the multiplier is small, often less than 1.0. That is, if the analyst allows only a very small probability of the links to exceed a very mild congestion, the network could not carry a large demand. This is in agreement with common sense. Similarly, toward the top-right region, the multiplier increases. This result permits the analyst to trade the congestion factor and travel time exceedance probability in order to accommodate a certain level of travel demand. For example, for the same demand multiplier of 1.5, at Point A in the top chart, if one specifies a tight exceedance probability of 0.1 (i.e, most of the links would not exceed a particular travel time), this travel time has to be 3 times the free-flow time. At point B, the same multiplier would result from $\beta_{ij} = 1.5, \phi_{ij} = 0.7$. That is, with the specified travel time of 1.5 times the free-flow travel time, on average each link has a 70% chance of exceeding this specified travel time.

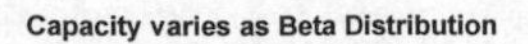

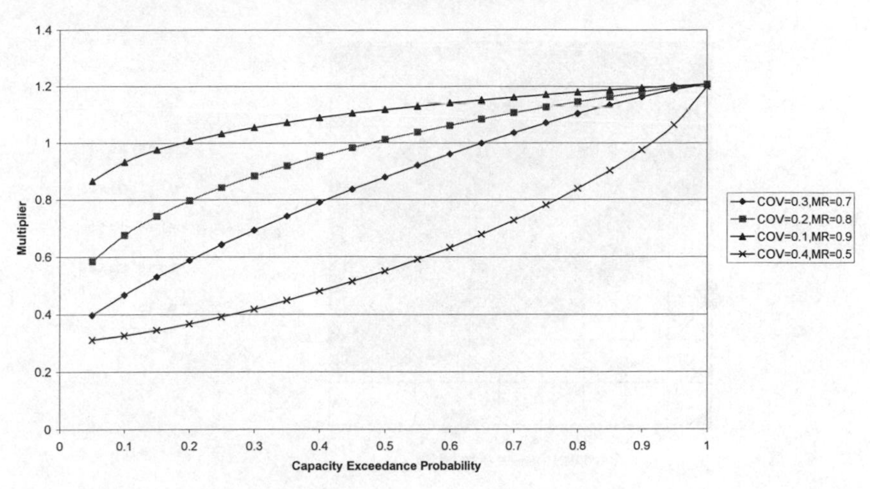

Figure 4 Demand multiplier with link capacity reliability.

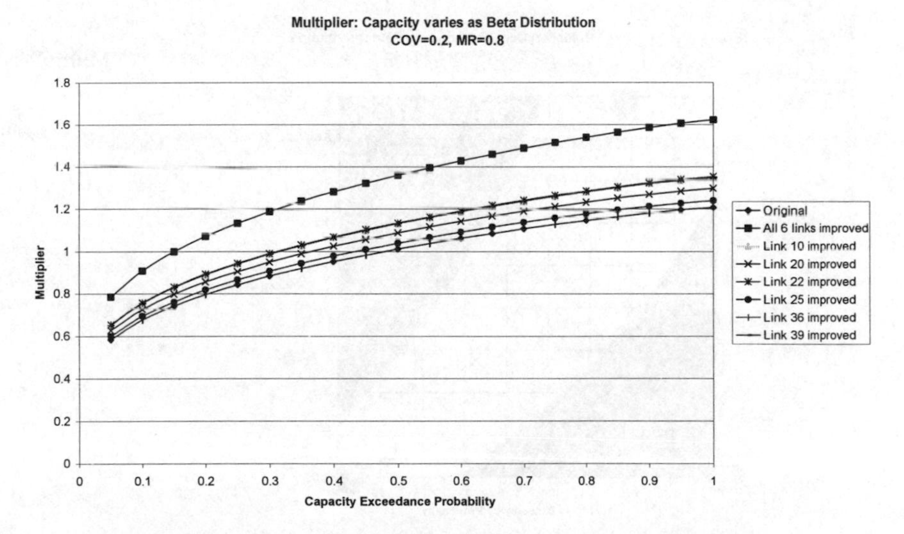

Figure 5 Demand multiplier with improved link capacities.

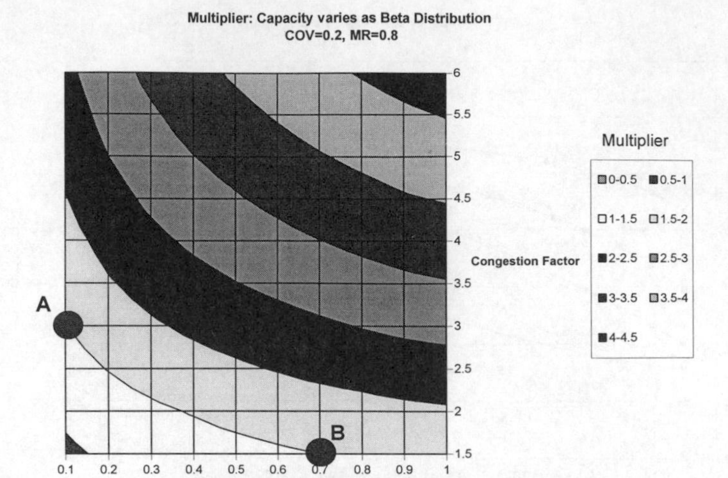

Multiplier

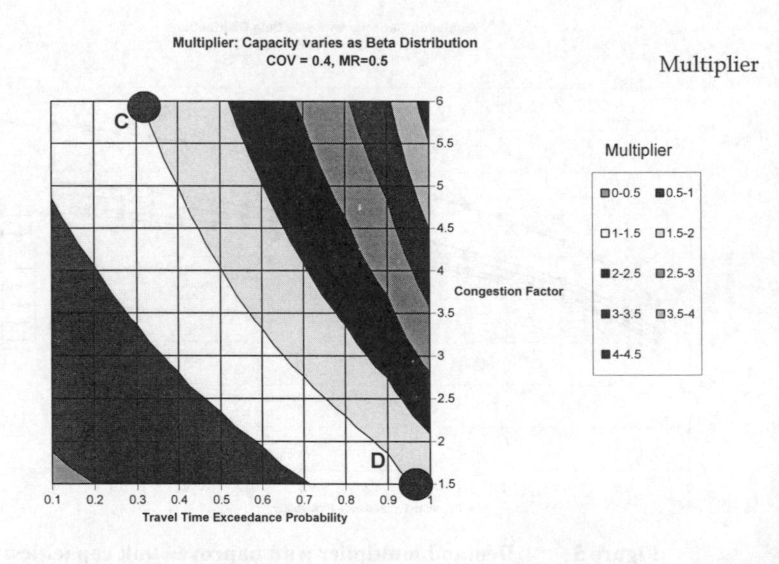

Figure 6 Demand multiplier with travel time reliability constraints.

In the top and bottom charts of Figure 6, one can see that for the same travel demand, the network with more severely degraded links (with a lower MR) has to suffer a much higher congestion factor and travel time exceedance probability. Examine the lines A-B and C-D in these charts, which have the same multiplier.

These two measures provide a very simple approach to analyzing network performance with degraded links. They can identify the links whose improvements would benefit the network most. Also, they provide a measure for the analyst to understand the relationship between network demand and congestion level in a stochastic manner.

4. CONCLUDING REMARKS

We have presented a chance constraint model for network capacity. Two types of reliabilities are considered in this study: link capacity and travel time reliabilities. By explicitly specifying the acceptable probability of exceeding each link's capacity or their specified travel time, we formulated a linear program to maximize the flow that can be carried by a network. This simple model provides two measures to understand the performance of a network with degraded links.

In this first model of this approach, we have not considered route choice behavior. This is the focus of an ongoing research. Such an extension is to be considered in a future study.

5. ACKNOWLEDGEMENTS

This research was sponsored by the research grant (RGC-DAG96/97.EG32) from the Hong Kong Research Grant Council.

REFERENCES

Aggarwal, K.K. (1985) Integration of reliability and capacity in performance measure of a telecommunication network. *IEEE Transactions on Reliability* 34, 184-186.

Ahuja, R.K., T.L. Magnanti, and J.B Orlin. (1993). *Network Flows: Theory, Algorithms, and Applications*, Prentice-Hall, Englewood Cliffs, NJ.

Ang, A.H.S. and W.H. Tang. (1990) *Probability concepts in engineering planning and design, vol.1 basic principle; vol. II – decision, risk and reliability*. John Wiley & Sons.

Asakura, Y. and M. Kashiwadani. (1991) Road network reliability caused by daily fluctuation of traffic flow. *Proceedings of the 19th PTRC Summer Annual Meeting*, Brighton, 73-84.

Asakura, Y. (1996) Reliability measures of an origin and destination pair in a deteriorated road network with variable flows. In: *Proceeding of the 4th Meeting of the EURO Working Group in Transportation.*

Bell, M.G.H. and Y. Iida. (1997) *Transportation Network Analysis*. John Wiley & Sons.

Bell, M.G.H., C. Cassir, Y. Iida, and W.H.K. Lam. (1999) A sensitivity based approach to reliability assessment. *Proceedings of the 14th ISTTT*, 283-300.

Billington, R. and W. Li. (1994) *Reliability Assessment of Electric Power Systems Using Monte Carlo Methods*. Plenum Press, New York.

Chan, Y. Yim, E., and A. Marsh. (1997) Exact & approximate improvement to the throughput of a stochastic network. *IEEE Transactions on Reliability* 46, 473-486.

Chen, A., H. Yang, H.K. Lo, and W.H. Tang. (1999a) A capacity related reliability for transportation networks. *Journal of Advanced Transportation*, 33(2), 183-200.

Chen, A., H. Yang, H. K. Lo, and W.H. Tang. (1999b). Capacity Reliability of a Road Network: An Assessment Methodology and Numerical Results. *Transportation Research*. In press.

Du, Z.P. and A. Nicholson. (1997) Degradable transportation systems: sensitivity and reliability analysis. *Transportation Research* 31B, 225-237.

Fuessle, R., D. Brill, and J. Liebman. (1987). Air quality planning: a general chance-constraint model. *Journal of Environmental Engineering*, ASCE, 113, 1, 106-123.

Guldmann, J. M. (1983). Supply, storage, and service reliability decisions by gas distribution utilities: a chance-constrained approach. *Management Science*, 29, 8, 884-906.

Iida, Y. and H. Wakabayashi. (1989) An approximation method of terminal reliability of a road network using partial minimal path and cut set. *Proceedings of the 5th WCTR*, Yokohama, 367-380.

Jacobs, T., M. Medina, and J. Ho. (1997). Chance constrained model for storm-water system design and rehabilitation. *Journal of Water Resources Planning and Management*, ASCE, 123, 3, 163-168.

LeBlanc, L, E. Morlok, and W. Pierskalla. (1975). An Efficient Approach to Solving the Road Network Equilibrium Traffic Assignment Problem. *Transportation Research*, 9, 430-442.

Li, D., T. Dolezal, and Y. Haimes. (1993) Capacity reliability of water distribution networks. *Reliability Engineering and System Safety,* 42, 29-38.

Mays, L. and Y.K. Tung. 1992. *Hydrosystems Engineering and Management*. McGraw-Hill, Inc. New York.

Wakabayashi, H. (1996) Reliability Assessment and importance analysis of highway network: a case study of the 1995 Kobe earthquake. *Proceedings of the First Conference of Hong Kong Society for Transportation Studies*, Hong Kong, 155-169.

Wong, S.C. and H. Yang. (1997) Reserve capacity of a signal-controlled road network. *Transportation Research* 31B, 397-402.

Yang, H. and M.G.H. Bell. (1998a) A Capacity Paradox in network design and how to avoid it. *Transportation Research*, 32A, 539-545.

Yang, H. and M.G.H. Bell. (1998b) Models and algorithms for road network design: A review and some new developments. *Transport Reviews*, 18, 257-278.

Chapter 12

Reliability Analysis and Calculation on Large Scale Transport Networks

Seungjae Lee
Assistant Professor
The University of Seoul
Email: sjlee@uoscc.uos.ac.kr

Byeongsup Moon
PhD Candidate
Seoul National University

Yasuo Asakura
Professor
Ehime University

1. Introduction

A driver tends to minimize his travel time. Traffic assignment models tend to emulate this travel behavior on the assumption that motorists choose the "shortest" path connecting an origin-destination pair. In the travel time based traffic assignment model the shortest path is defined and calculated on the basis of travel time on links. In some situations, the travel time based traffic assignment model does not emulate accurately driver's route choice behavior. Motorists do not choose the path according to only the travel time in some cases. In reality they have a tendency to prefer the path on which the travel time variation is minimized.

Network flows are influenced by abnormal or incident events that affect network characteristics and capacity such as disasters, accidents, construction or repair. Traffic congestion gives rise to variations in driver's behavior. Therefore, motorists tend to choose the path on which travel time variation is minimized.

In this chapter, in order to include the various kinds of travel behavior into the assignment model, a reliability traffic assignment model is developed. In systems engineering, reliability may be defined as the degree of stability in the quality of service which a system normally offers. Reliability analysis for transportation networks differs from general system reliability analysis, in that path choice behavior must be considered. We use the concept of reliability to reflect the driver's appreciation of path travel time variations and assume that a driver chooses the path according to the reliability.

This chapter defines the reliability based route choice principle and formulates a reliability based equilibrium traffic assignment using this principle. Reliability is defined as the difference of travel demand and capacity using the interference theory of system engineering. An efficient solution algorithm based on the Frank-Wolfe algorithm is presented to calculate and compare the reliability based traffic assignment with conventional travel time based assignment using small, medium and large scale road networks.

In the chapter, section 2 reviews some previous research on reliability in transportation. In section 3, travel time reliability and reliability user equilibrium is defined. The reliability assignment model is formulated as equivalent reliability user equilibrium and then it is demonstrated that the solution of which is revealed in reliability user equilibrium flow pattern. In section 4, the reliability traffic assignment model is applied to simple, middle and large scale networks. Section 5 is then dedicated to the results of the reliability assignment model.

2. BACKGROUND

Polus and Schofer (1976) researched the reliability of urban freeway operation. The measure of performance they used is lane occupancy. Reliability is defined as the operational consistency of a facility over an extended period of time. Using lane occupancy as the principal measure, consistency is assessed by comparing daily measures of occupancy, for many days, in order to capture the major variational trends in a facility performance.

Under ideal conditions, one might expect all observed values to be about the same and thus the distribution would have a small variance. In this case, the facility could be described as highly reliable. As the variability in observed occupancy measures increases, it may be inferred that the freeway facility is less consistent in its operation, and thus of lower reliability. Based on the foregoing logic, freeway reliability is viewed in terms of the inverse of its performance variability.

Ferrari (1988) presented an analysis of motorway circulation under congested traffic conditions, with the aim of arriving at a measurement of the reliability of the motorway transport system. He suggested a model of the speed process on a motorway lane under conditions of congestion, which was tested experimentally employing a great number of data collected on two motorways, varying considerably in terms of environmental and traffic conditions. Reliability is defined by the probability that, within a certain period of time starting from the instant when the reliability is measured, no drops in speed to an extent deemed hazardous will occur.

Wakabayashi and Iida (1994) defined that transportation network reliability has two aspects: connectivity and travel time reliability. Connectivity is the probability that traffic can reach a given destination at all, while travel time reliability is the probability that traffic can reach a given destination within a given time. They propose a heuristic method, which calculates the system reliability through the Esary-Proschan upper and lower bound using Boolean algebra. But the heuristic method calculates the basic idea of transportation network reliability in terms of connectivity only, and not travel time.

Du and Nicholson (1993) proposed a new approximation method for analyzing transportation network reliability, taking into account the flow characteristics in addition to the topological structure of the network. This method makes full use of information about a number of the most probable component state vectors (link capacity vectors), instead of solving the integrated equilibrium model of a degradable transportation system for all possible component state vectors. Using this algorithm, the connectivity reliabilities of a degradable transportation system can be determined with reasonable computation effort.

Willumsen and Hounsell (1998) stated that a key factor affecting travel times and their variability is the level of congestion, which itself depends on the relationship between demand and supply. Factors which influence demand include time of day, day of the week, and time of year. Factors which influence supply include weather conditions, lighting conditions, and incident such as accidents, breakdowns, signal failures and road works. They chose the standard deviation of travel time as the main indicator of travel time variability, and they used free flow travel time, delay and a congestion index as explanatory variables. Regression relationships were investigated between travel time variability and the proposed explanatory variables. The travel time variation is affected by not one factor but multiple factors. A new measurement including all these factors is needed. In this research they propose the relationship between road capacity and traffic flow as the measurement of travel time variation and travel time reliability.

3. RELIABILITY ASSIGNMENT MODEL

3.1 Interference theory

We introduce the interference theory in reliability design for defining the reliability function. In interference theory a system fails when the stress exceeds the strength. Let the density function for the stress be denoted by s and that for strength by S, as shown in Figure 1. Then by definition,

$$\text{Reliability} = R = P(S > s) = P(S - s > 0)$$

The shaded portion in Figure 1 shows the interference area, which is indicative of the probability of failure. Let us enlarge this, as shown in Figure 2, in order to better focus our attention on it. The probability of a stress value lying in a small interval of width *ds* is equal to the area of the element *ds* ; that is,

$$p(s_0 - \frac{ds}{2} < s < s_0 + \frac{ds}{2}) = f_s(s_o) \cdot ds$$

The probability that the strength S is greater than a certain stress s_0 is given by

$$P(S > s_0) = \int_{s_0}^{\infty} f_S(S)\, dS$$

The probability for a stress value lying in the small interval *ds* and the strength S exceeding the stress given by this small interval *ds* under the assumption that the stress and the strength random variables are independent is given by

$$f_s(s_0)\,ds \cdot \int_{s_0}^{\infty} f_S(S)\,dS$$

The reliability of the component is the probability that the strength S is greater than the stress s for all possible values of the stress s and hence is given by

$$R = \int_{-\infty}^{\infty} f_s(s)\left[\int_{s}^{\infty} f_S(S)\,dS\right]ds \tag{1}$$

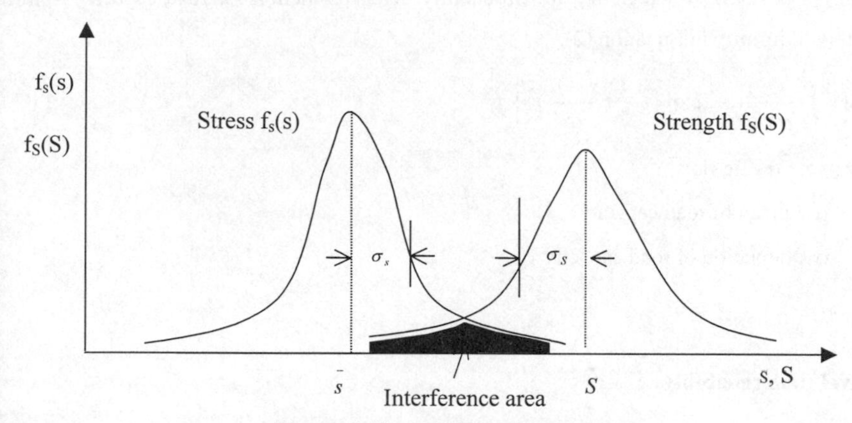

Figure 1. Stress-strength Interference.

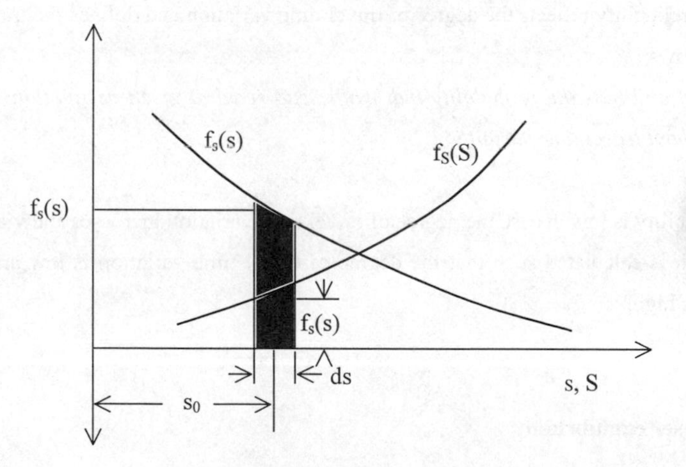

Figure 2. Computation of Reliability.

Stress and strength in interference theory can be seen as the same as traffic flow and road capacity in transportation. When stress is greater than strength in interference, the device fails. In the same manner, when traffic flow is greater than road capacity, the probability of traffic congestion and travel time variations increases.

We assume that traffic flow is a deterministic value since we use the All-or-Nothing method in the reliability subroutine assignment model and road capacity has a normal distribution because capacity is stochastically affected by incident or road condition. Therefore, reliability in Equation (1) is given by integrating the probability density function for road capacity from a traffic flow to infinity in Equation (2).

$$R(x) = \int_x^\infty \frac{1}{\sigma\sqrt{(2\pi)}} \exp\left\{-\frac{1}{2}\left(\frac{s-\mu}{\sigma}\right)^2\right\} ds \qquad (2)$$

where x : traffic flow

μ : mean of road capacity

σ : deviation of road capacity

3.2 Travel time reliability

We assume that reliability reflects the degree of travel time variation and defines the travel time reliability as follows:

Travel time reliability is the probability that traffic can reach a given destination within a given time without travel time variation.

Travel time reliability is low in case the degree of travel time variation increases and vice versa. The shortest path is calculated such that the degree of travel time variation is low and travel time reliability is high.

3.3 Reliability user equilibrium

To solve the reliability assignment model, it is required that the rule by which motorists choose a route be specified. The interaction between the routes chosen between all O-D pairs, on the one hand, and the travel time reliability on all the network links, on the other, determines the equilibrium flows.

It is reasonable to assume that every motorist will try to minimize his or her own travel time variation when traveling from origin to destination. This does not mean that all travelers between each origin and destination pair should be assigned to a single path. The travel time variation on each link changes with the flow and therefore, the travel time reliability on several of the network paths changes as the link flows change. A stable condition is reached only when no traveler can improve his travel time reliability by unilaterally changing routes. This is the characterization of the reliability user equilibrium.

3.4 Formulating the reliability assignment problem

To demonstrate how the reliability user equilibrium flow pattern can be found for a large network, we use an equivalent minimization method. This approach involves the formulation of a mathematical program, the solution of which is revealed in the reliability user equilibrium flow pattern. This approach is used often in operations research, in cases where it is easier to minimize the equivalent program than to solve a set of conditions directly.

In a transportation network a path consists of a series of links. So, the reliability of each path is calculated in Equation (3). And Equation (3) is converted to Equation (4) through the logarithm function (see Bell and Iida 1997).

$$R_A = R_1 \times R_2 \times \bullet\bullet\bullet \times R_n = \prod_i R_i \quad i = 1,2,\ldots,n, \ i \in A \tag{3}$$

$$\ln(R_A) = \ln(\prod_i R_i) = \sum_i \ln(R_i) \tag{4}$$

The reliability equilibrium assignment problem is to find the link flows, x, that satisfy the reliability user equilibrium criterion when all the origin-destination entries, q, have been appropriately assigned. This link-flow pattern can be obtained by solving the following mathematical program:

$$\text{MAX } z(x) = \sum_a \int_0^{x_a} \ln R_a(w)\,dw \tag{5}$$

Equation (5) can be converted to the minimization problem in Equation (6).

$$\text{MIN } z(x) = -\sum_a \int_0^{x_a} \ln R_a(w)\,dw \tag{6}$$

$$\text{s.t.} \quad \sum_k f_k^{rs} = q_{rs} \quad \forall \text{ r, s} \tag{6-1}$$

$$f_k^{rs} \geq 0 \quad \forall \text{ k, r, s} \tag{6-2}$$

$$x_0 = \sum_r \sum_s \sum_k f_k^{rs} g_{k,a}^{rs} \quad \forall_a \tag{6-3}$$

$$R_a(w) = \int_w^{\infty} \frac{1}{\sigma\sqrt{(2\pi)}} \exp\left\{-\frac{1}{2}(\frac{x-\mu}{\sigma})^2\right\} dx \tag{6-4}$$

where x_a : flow on link a

f_k^{rs} : flow on path k connecting O-D pair r-s

q_{rs} : trip rate between origin r and destination s

$\delta_{a,k}^{rs}$: indicator variable

$R_a(\omega)$: travel time reliability on link a in case traffic flow is ω

In this formulation, the objective function is the sum of the integrals of the travel time reliability. Eq.(6-1) represents a set of flow conservation constraints. The nonnegativity conditions in Eq.(6-2) are required to ensure that the solution of the program will be physically meaningful. Eq.(6-3) represents the definitional link-path incidence relationships and expresses the link flows in terms of the path flows. Eq.(6-4) is the travel time reliability on link a.

3.5 Equivalency condition

To demonstrate the equivalence of the equilibrium assignment problem and program (6), it has to be shown that any flow pattern that solves (6) also satisfies the equilibrium conditions. This equivalency is demonstrated by proving that the first-order conditions for the minimization program are identical to the equilibrium conditions.

The Lagrangian of the equivalent minimization problem with respect to the equality constraints (6-1) can be formulated as

$$\text{L(f,u)} = z[x(f)] + \sum_{rs} u_{rs}(q_{rs} - \sum_k f_k^{rs}) \tag{7}$$

The first-order conditions of program (6) are equivalent to the first-conditions of Lagrangian

(7). At the stationary point of the Lagrangian, the following conditions have to hold with respect to the path-flow variables:

$$f_k^{rs}\frac{\partial L(f,u)}{\partial f_k^{rs}}=0\ ,\quad \frac{\partial L(f,u)}{\partial f_k^{rs}}\geq 0 \quad \forall \quad k,\ r,\ s \tag{7-1}$$

$$\frac{\partial L(f,u)}{\partial u_{rs}}=0 \quad \forall\ r\ ,\ s \tag{7-2}$$

Also, the nonnegativity constraints have to hold. Condition (7-2) has to hold at equilibrium. The first-order conditions expressed in Eq.(7-1) can be obtained explicitly by calculating the partial derivatives of L(f,u) with respect to the flow variables, f_l^{mn}, and substituting the result into (7-1). This derivative is given by

$$\frac{\partial}{\partial f_l^{mn}}L(f,u)=\frac{\partial}{\partial f_l^{mn}}z[x(f)] \quad + \quad \frac{\partial}{\partial f_l^{mn}}\sum_{rs}u_{rs}(q_{rs}-\sum_k f_k^{rs}) \tag{8}$$

The first term on the right-hand side of Eq.(8) is the derivative of the objective function (6) with respect to f_l^{mn}. This derivative can be evaluated by using the chain rule :

$$\frac{\partial z[x(f)]}{\partial f_l^{mn}}=\sum_{b\in A}\frac{\partial z(x)}{\partial x_b}\frac{\partial x_b}{\partial f_l^{mn}} \tag{9}$$

The first quantity in the r.h.s. of Eq.(9) can be easily calculated since the travel time reliability on any link is a function of the flow on that link only.

$$\frac{\partial z(x)}{\partial x_b}=\frac{\partial}{\partial x_b}\left\{-\sum_a\int_0^{x_a}\ln R_a(\omega)d\omega\right\}=-\ln R_b \tag{10}$$

The second quantity in the product is the partial derivative of a link flow with respect to the flow on a particular path.

$$\frac{\partial x_b}{\partial f_l^{mn}}=\delta_{b,l}^{mn} \tag{11}$$

Substituting the last two expressions into Eq.(9), the derivative of the objective function with respect to the flow on a particular path becomes

$$\frac{\partial z[x(f)]}{\partial f_l^{mn}}=\sum_b -\ln R_b\delta_{b,l}^{mn}=c_l^{mn} \tag{12}$$

In other words, it is the travel time reliability on that particular path.

The second type of term in Eq.(8) is even simpler to calculate.

$$\frac{\partial f}{\partial f_l^{mn}} \sum_{rs} u_{rs}(q_{rs} - \sum_k f_k^{rs}) = -u_{mn} \tag{13}$$

Substituting both (12) and (13) into Eq.(8), the partial derivative of the Lagrangian becomes

$$\frac{\partial}{\partial f_l^{mn}} L(f,u) = c_l^{mn} - u_{mn} \tag{14}$$

The general first-order conditions for the minimization program in Eqs.(6) can now be expressed explicitly as

$$f_k^{rs}(c_k^{rs} - u_{rs}) = 0 \quad \forall\ k,\ r,s \tag{15-1}$$

$$c_k^{rs} - u_{rs} \geq 0 \quad \forall\ k,\ r,\ s \tag{15-2}$$

$$\sum_k f_k^{rs} = q_{rs} \quad \forall\ r,\ s \tag{15-3}$$

$$f_k^{rs} \geq 0 \quad \forall\ k,\ r,\ s \tag{15-4}$$

Conditions (15-3) and (15-4) hold at the point that minimizes the objective function. For a given path, say path *k* connecting origin r destination s, conditions (15-1) and (15-2) hold for two possible combinations of path flow and travel time reliability. Either the flow on that path is zero, in which case the logarithm of travel time reliability on this path, c_k^{rs}, must be less than or equal to the O-D specific Lagrange multiplier, u_{rs}; or the flow on the *k* th path is positive, in which case $c_k^{rs} = u_{rs}$, which means that the reliability of the path is equal to $\exp(u_{rs})$ and that no paths with lower reliability are used and both Eqs.(15-1) and (15-2) hold as equalities. Thus $\exp(u_{rs})$ equals the minimum path travel time reliability between origin r and destination s. With this interpretation, it is now clear that Eqs.(15), in fact, state the reliability user equilibrium principle. Accordingly, this program is referred to as the reliability user equilibrium program.

It seems sufficient if we assume any probability distribution of link capacity. The reliability function R(x) is always a decreasing function for any functional form of the p.d.f. of link capacity. Then -ln (R(x)) is always an increasing function for x (link flow). The convexity of the objective function is then satisfied for any p.d.f. of link capacity.

3.6 Solution algorithm

The algorithm itself, when applied to the solution of the reliability user equilibrium problem, can be summarized as follows:

Step 0 : Initialization. Perform all-or-nothing assignment based on

$R_a = R_a(0), \forall a$. This yields $\{x_a^1\}$. Set counter n = 1.

Step 1 : Update. Set $R_a^n = R_a(x_a^n), \forall a$.

Step 2 : Direction finding. Perform all-or-nothing assignment based on R_a^n.

This yields a set of (auxiliary) flows $\{y_a^n\}$.

Step 3 : Line search. Find α_n that solves

$$\underset{0 \le \alpha \le 1}{MIN} - \sum_a \int_0^{x_a^n + \alpha(y_a^n - x_a^n)} \ln R_a(\omega)\,d\omega$$

Step 4 : Move. Set $x_a^{n+1} = x_a^n + \alpha_n(y_a^n - x_a^n), \forall a$.

Step 5 : Convergence test. If a convergence criterion is met, stop (the current solution, $\{x_a^{n+1}\}$, is the set of equilibrium link flows) ;

otherwise, set n = n + 1 and go to step 1.

4. NUMERICAL EXAMPLES

4.1 Simple network

To demonstrate the convergence of the reliability assignment model to the equilibrium solution, it is applied to a simple network in Figure 3. Network has three links and one O-D pair. The link itself is the path. The characteristic of the link is depicted in Table 1. The deviation of road capacity is assumed to be half of the mean of road capacity. And O-D trip rates are 10 vehicles.

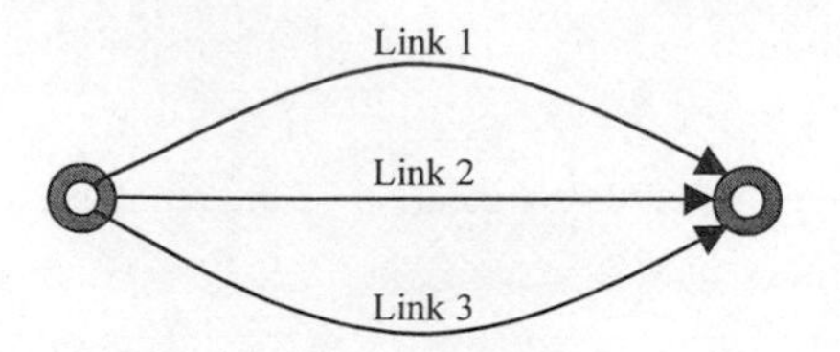

Figure 3. Simple Network.

Table 1. Link Characteristics.

Link	Mean of Road capacity	Deviation of road capacity	Initial Flow	Free flow Travel time
1	2	1	0	10
2	4	2	0	20
3	3	1.5	0	25

The results of the conventional assignment model where the BPR function is used are shown in Table 2. From Table 2 it is evident that after 6 iterations the flows are close to equilibrium ; the travel times on all three routes are very similar.

Table 2. Results of Travel Time Assignment Model.

Iteration number	Travel time			Traffic flow			Step size
	Link 1	Link 2	Link 3	Link 1	Link 2	Link 3	
0	10.0	20.0	25.0	10.00	0.00	0.00	-
1	947.5	20.0	25.0	4.03	5.97	0.00	0.5965
2	34.8	34.8	25.0	3.38	5.00	1.61	0.1611
3	22.3	27.4	25.3	3.62	4.83	1.55	0.0356
4	26.1	26.4	25.3	3.55	4.73	1.73	0.0204
5	24.8	25.9	25.4	3.59	4.69	1.71	0.0072
6	25.6	25.7	25.4	3.57	4.67	1.76	0.0054

The reliability assignment model is applied to the simple network in Figure 3. The results are shown in Table 3.

Table 3. Results of Reliability Assignment Model.

Iteration number	Travel time reliability			Traffic flow			Step Size
	Link 1	Link 2	Link 3	Link 1	Link 2	Link 3	
0	34.65	0.02	0.02	0.0	10.0	0.0	-
1	0.02	6.61	0.02	0.0	5.7	4.3	0.42855
2	0.02	1.63	1.63	2.2	4.4	3.3	0.22213
3	0.89	0.89	0.89	2.2	4.4	3.3	0.00001
4	0.89	0.89	0.89	2.2	4.4	3.3	0.00000

From Table 3 it is evident that after 4 iterations the flows are close to reliability user equilibrium; the travel time reliability on all three routes is the same. A comparison of this reliability model with the previous model can be made on the basis of the flow pattern at the end of iterations. In the travel time assignment model, traffic flow of link 2 is greatest because the capacity on link 2 is greatest. And traffic flow of link 1 is more than link 3. This is interpreted from the difference in free flow travel time.

In the reliability assignment model, traffic flow of link 2 is greatest as in the travel time assignment model. But traffic flow of link 3 is more than link 1. It may be inferred from these results that traffic is assigned according to the road capacity in the reliability assignment model.

4.2 Middle network

The reliability assignment model is applied to the Sioux-Fall network. The Sioux-Fall is constituted of 24 nodes, 24 zones and 76 links. The relative gap function used as convergence criterion in the reliability model is shown in Figure 4. As iterations increase, the function converges to 0. Therefore, it can be inferred that it is in equilibrium after 31 iterations.

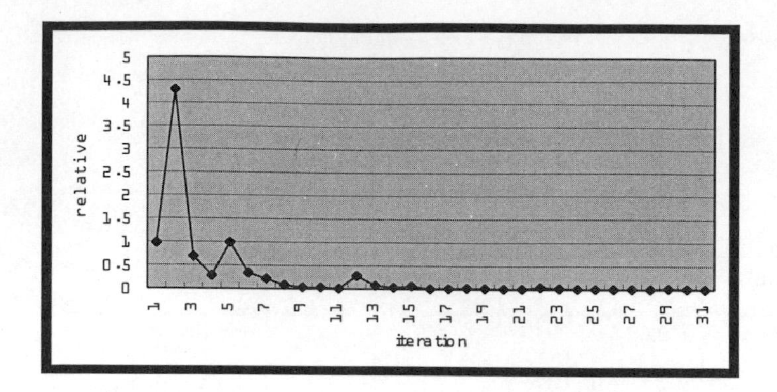

Figure 4. Relative Gap Function.

In order to compare the reliability model with the travel time model applied to the Sioux-Fall network, traffic flows of both link 1(origin 1 - destination 2) and link 2 (origin 1 -destination 3) are depicted in Figure 5. After 40 iterations, the travel time assignment model is in equilibrium. Each link flow has converged uniquely as the iterations end. Traffic flows of two links in both models are different because of the difference of the link performance function included.

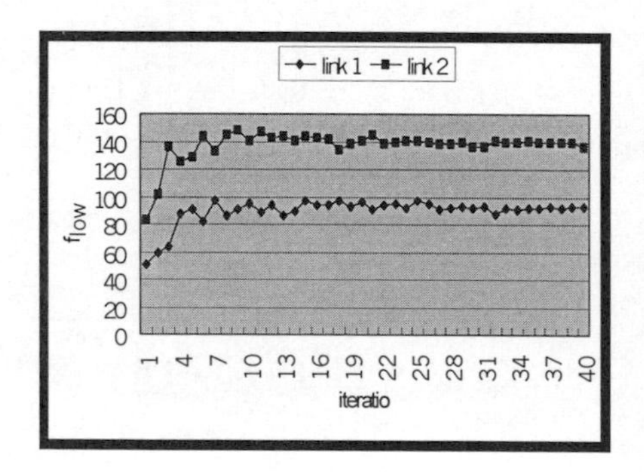

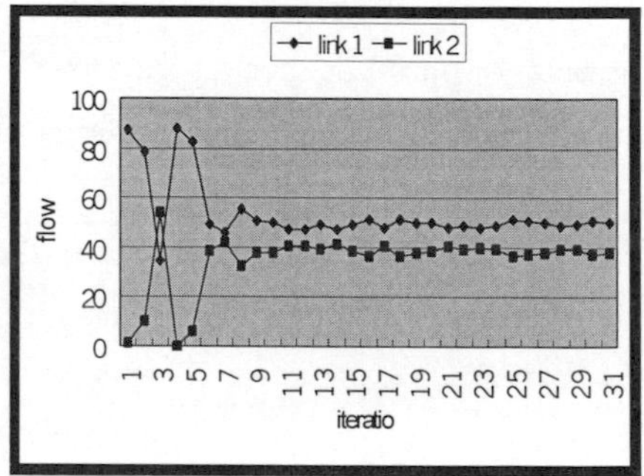

(a) travel time model (b) reliability model

Figure 5. Traffic Flows of Link 1 and Link 2.

4.3 Large network

The reliability assignment model is also applied to the network of the city of Ansan, Korea. The network of Ansan city is constituted of 672 nodes and 2224 links. We divide the network into 20 zones on the basis of administrative district. And traffic departing from the centroid of each zone is assigned according to both the reliability assignment model and the travel time model.

After 21 iterations the travel time model is in equilibrium and after 18 the reliability model is also in equilibrium. The relative gap function in the reliability model converged to 0 in Figure 6. Therefore, it is demonstrated that the reliability assignment model brings the large network toward reliability equilibrium.

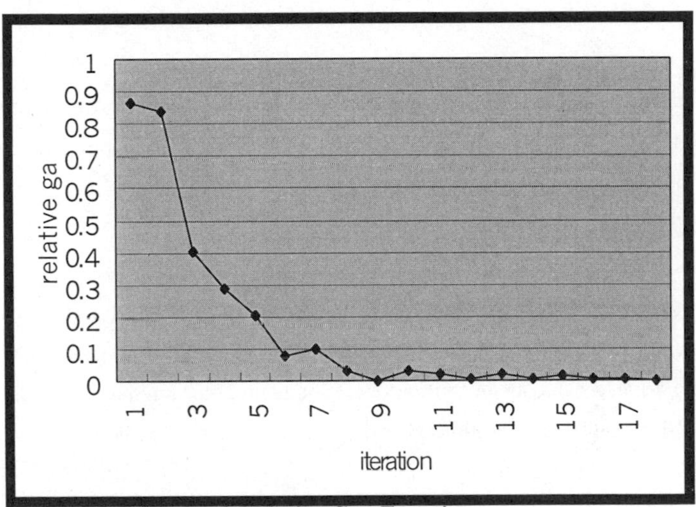

Figure 6. Relative Gap Function.

For comparing the reliability model with the travel time model, traffic flows of both link 1(node 1 - node 2) and link 2(node 1 - node 3) are depicted in Figure 7. Node 1 is the centroid of the 1st zone. Each link flow converged uniquely as iteration ends.

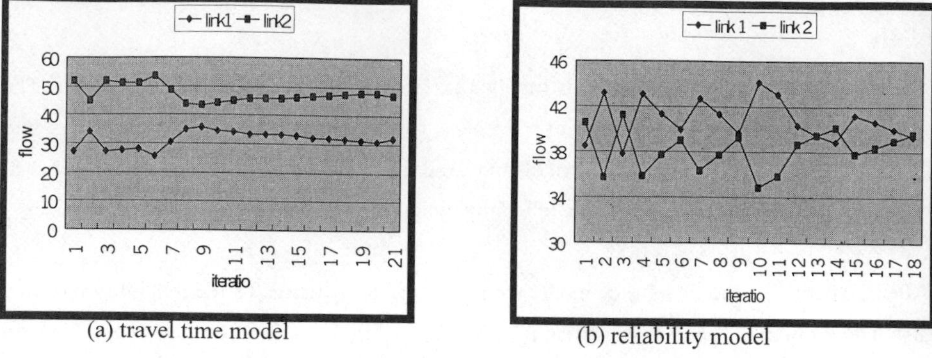

(a) travel time model (b) reliability model

Figure 7. Traffic Flows of Link 1 and Link 2.

5. CONCLUSIONS

The effectiveness of any traffic assignment model depends on how accurately the model emulates the real traffic pattern. Driver's behavior in traffic congestion is different from that in uncongested traffic conditions. When traffic is uncongested, motorists choose the path on the basis of travel time. But in traffic congestion, motorists might choose the path considering travel time variation caused by incidents, weather conditions or traffic conditions.

We formulate a reliability assignment model that includes the travel time reliability in order to reflect the driver's behavior in traffic congestion. Travel time reliability is determined by the degree of the travel time variation on paths which the motorists choose. Travel time variation occurs through the relationship between road capacity and traffic flow. So, travel time reliability is a function of both road capacity and traffic demand.

The reliability assignment model presented in this chapter was shown to satisfy some reliability user equilibrium conditions through mathematical proof and sample networks. In this research, the reliability assignment model is applied to test networks on which imaginary traffic flows are assigned. In future research the model will be applied to real networks with field data. For use in traffic management, a dynamic reliability assignment model may be developed.

6. REFERENCES

Bell, M. and Y. Iida (1997) Transport network analysis. John Wiley & Sons, 120-135.

Billinton, R. and R.N. Allan (1992) Reliability evaluation of engineering systems. Second edition, Plenum Press, 19-92.

Brunilde S., and F. Soumis (1991) Communication & transportation network reliability using routing models. *IEEE transactions on reliability*, Vol. 40, No. 1, 29-38.

Castillo, E. (1988) Extreme value theory in engineering, Boston : Academic, 87-98.

Dhillon, B. S. (1983) Reliability engineering in systems design and operation. Van Nostrand Reinhold Company, 45-50.

Hiroshi,W. and Y. Iida (1994) Improvement of road network reliability with traffic management. *IFAC transportation systems*, 603-608

Kapur, K. and L.R. Lamberson (1977) Reliability in engineering design. John Wiley & Sons, 19-56.

Martz, H. F. and A. Ray (1982) Bayesian reliability analysis. Network : Wiley, 1982, 34-50.

Moon, B., S. Lee and K. Lim (1997) Reliability assignment considering driver's behavior under an emergency condition. *Proceedings of the 4th World Congress on Intelligent Transport Systems*. 3-6.

Nicholson, A. and Z. Du (1997) Degradable transportation systems: sensitivity and reliability analysis. *Transportation Research, B*, Vol. 31, No. 3, 225-237

Nicholson, A. and Z. Du (1997). Degradable transportation systems: an integrated equilibrium model. *Transportation Research B*, Vol. 31, No. 3, 209-223.

Rao, S. (1992) Reliability-based design. McGraw-hill, Inc., 79-92.

CHAPTER 13

ROUTE CHOICE TO MAXIMISE THE PROBABILITY OF ARRIVAL WITHIN A SPECIFIED TIME

Mike Maher and Xiaoyan Zhang
Transport Research Institute, Napier University, Edinburgh, Scotland

1. INTRODUCTION

Travel times on road links are subject to day-to-day variation and therefore, recognising this variability, a driver may in some circumstances wish to choose a route which does not necessarily minimise expected journey time but instead maximises the probability of arrival at the destination within a specified time. This chapter proposes a new route choice model using this probability of arrival within a specified time as the objective function and shows that the problem can be formulated as a many-to-one, Markov process, dynamic programming problem. This "dynamic" version of the route choice problem (in which, at each node, drivers select their optimal exit link, depending upon the time remaining) is contrasted with the "static" problem in which drivers select their routes at the start of their journeys. If the profile of drivers' "target" journey times is known, the expected link volumes can also be obtained. The solution algorithm for the dynamic problem is link-based and hence avoids the need for path enumeration. To illustrate its operation, the algorithm is applied to a small network. The chapter goes on to show that the algorithm can also be applied with a more general form of utility function, and suggests how it may be modified to enable it to deal with networks containing loops.

2. Structure of the Chapter

In this chapter we address the problem of travel time reliability in a road network, and define travel time reliability as the probability that a trip can reach its destination within a specified time, as in Bell and Iida (1997). Travel times on road links (and hence on routes) are subject to day-to-day variation. A driver's choice of route may be affected not only by the expected journey time but also (or instead) by its reliability. Given an urgent deadline (such as a meeting or a lecture), he may choose the route that offers maximum reliability rather than minimum expected travel time.

After introducing the notation and formulating the problem in section 3, a number of alternative route choice criteria are proposed for the case in which the driver selects his route at the start of his journey: the "static" route choice problem. Because of the difficulties associated with path enumeration in large networks, we then formulate in section 4 a "dynamic" version of the problem, in which, at each node, the driver selects an exit link. It is then shown how this can be solved efficiently by a many-to-one, Markov process, dynamic programming algorithm and how, given data on the distribution of tripmakers' "target" journey times, the expected traffic volume on each link can be determined. The application of this algorithm to a small grid network is demonstrated in section 5. It is then shown in section 6 that a more generalised form of objective function, which allows for other definitions of reliability, can be adopted without any alteration to the structure of the algorithm. Finally, section 7 draws conclusions and indicates how the approach may be used to quantify the network reliability benefits from a highway construction or improvement scheme.

3. Formulation of the Problem

We assume that each link (i, j) has a journey time distribution, which describes the day-to-day variation, and which has probability function $p_{ij}(t)$. Note that we are using a discrete time formulation. Particular features of these distributions include, for example, the minimum, mean and maximum values t_{min}, t_{av}, and t_{max}.

As any route is defined as a sequence of links, it follows that the journey time distribution for any route can be found from the convolution of the link journey time distributions. For example, the probability function of the journey time along the path made up of two successive links (i, j) and (j, k) is given by:

$$p_{ijk}(t) = \sum_s p_{ij}(s) p_{jk}(t-s) \tag{1}$$

Hence, by repeated application of (1), the journey time distribution for any path $(i, j, k, l \ldots)$ can be found.

A number of alternative route choice criteria can be proposed. For example, an optimist might choose the route which offers the minimum possible journey time, whilst a realist might choose that which offers the minimum expected journey time, and a pessimist might select that which minimises the maximum journey time. In each case, the optimal route can be found by means of a conventional shortest path algorithm (such as that of Dijkstra, 1959), with the link costs being defined, respectively, as t_{min}, t_{av}, and t_{max}.

However, suppose that the driver is interested in choosing the most reliable route, where reliability is defined as the probability $F_r(t)$ of arriving at his destination, node n, within a specified time t, using route r. Then he should compare the $F_r(t)$ for each possible route r and select that which is the maximum, given his "target" travel time t. Figure 1 illustrates a simple case with just three routes, from which it may be seen that if $t < 41$ mins, the driver should select route 3, whilst if $41 < t < 45$ he should select route 2, and if $t > 45$ he should select route 1. However, in even moderately-sized networks, it is not feasible to enumerate all possible routes. Therefore, the aim here is to formulate the problem in such a way as to avoid the need for path enumeration, by adopting a link-based approach.

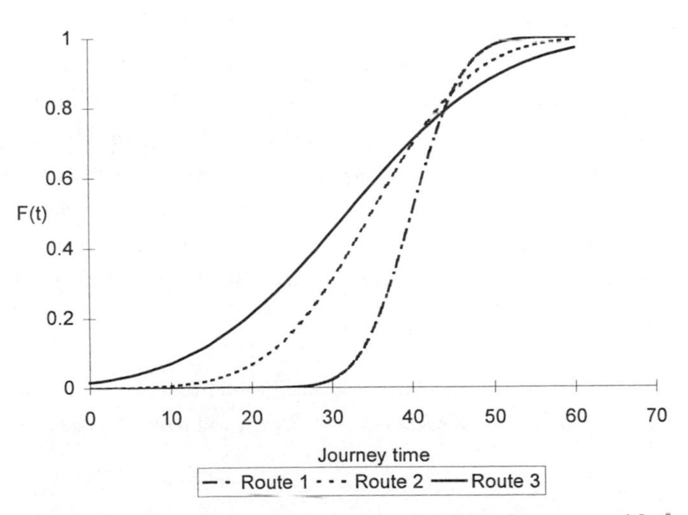

Figure 1 : route choice to maximise reliability, for a case with three routes.

4. A Dynamic Route Choice Algorithm

Instead of the "static" route choice problem considered in the previous section, under which a driver selects his most reliable route at the start of his journey and does not deviate from it, whatever might happen during the journey, we now formulate a "dynamic" version in which the driver need only decide at each node which exit link to take. This decision may depend on the time he has remaining to reach his destination.

The problem will be formulated as a "many-to-one" (that is, many origins to one destination, node n), Markov process, dynamic programming problem. First, we define $F_k{}^*(t)$ as the *maximum* probability of arriving at n, starting from node k, with time t remaining. Note that this assumes an optimal choice of exit link at each intermediate node between k and n.

Imagine the driver is currently at node k, with time t remaining, and must select an exit link (or equivalently, a successor node l). For each such exit link, there is a probability $p_{kl}(s)$ of a travel time s, with the consequence that, at node l, the remaining time will be $(t - s)$. Assuming that the driver chooses his route in an optimal manner thereafter, the following system of equations can be written:

$$F_k{}^*(t) = \max_l \sum_s p_{kl}(s) F_l{}^*(t-s), \quad \forall k, t \tag{2}$$

where the maximisation for l is over the set of successor nodes for node k and the summation for s is from the minimum to the maximum travel time for link (k, l). This is a dynamic programming problem (see, for example, Bellman 1957 and Howard 1960), as it is solved in a sequence of stages, working backwards through the network, calculating the $F_k{}^*(t)$ for one node at a time, starting from the destination, where we note that $F_n{}^*(t) = 0$ for $t < 0$ and $= 1$ for $t \geq 0$. (We shall assume for the moment that the network does not contain any loops, so that a sequence of nodes exists for which the equations above can be solved recursively. We shall return to this point later.) The formulation is Markovian as the choice of optimal exit link depends only on a knowledge of the current state, defined as the current node k and the time remaining t, and not on any previous history. This relies on the assumption of independence of link travel times.

At the end of the algorithm, $F_k{}^*(t)$ will have been calculated for all feasible values of t, for all nodes k. In addition, the optimal choice of successor node $L_k{}^*(t)$ will have also been calculated and stored for all states (k, t). (If there is more than one destination node, the algorithm must be applied separately for each one.)

Given a knowledge of the number of drivers $Q_m(t)$ travelling from origin m to the destination node n with "target" journey time t, it is then possible to calculate the expected numbers using each link. Let $q_k(t)$ denote the number of drivers entering node k (having origins before k) with time t remaining to arrive at destination node n. Then:

$$q_l(t) = \sum_k \sum_s (Q_k(t+s) + q_k(t+s)) p_{kl}(s) \delta_{kl}(t+s), \quad \forall l, t \tag{3}$$

where $\delta_{kl}(t) = 1$ for $l = L_k^*(t)$ and zero otherwise. The $q_l(t)$ can be calculated in the reverse order of nodes to that employed in the calculation of the $F_k^*(t)$ (that is, a forward pass through the network rather than a backward pass). The total volume v_{kl} on link (k, l) is then given by

$$v_{kl} = \sum_t \sum_s (Q_k(t+s) + q_k(t+s)) p_{kl}(s) \delta_{kl}(t+s) \tag{4}$$

It should be noted that dynamic routeing is bound to produce results which are at least as good as static routeing. That is, the value of $F_m^*(t)$ is certainly no less than the maximum of the $F_r(t)$ over all the static routes r between any origin m and destination n.

5. APPLICATION OF THE ALGORITHM TO A SMALL TEST NETWORK

Consider now the application of the algorithm described in section 4 to a small test network, consisting of a twelve link, nine node grid as shown in Figure 2. The links in the network are of three types: type A has a small value of minimum travel time but is quite variable, type C is the opposite, having a higher minimum travel time but being much less variable, and type B is medium in both respects. The probability functions of their travel times $p(t)$ are shown in Table 1.

Table 1 : probability function $p(t)$ for each of the three link types A, B and C.

t	1	2	3	4	5	6	7	8
$p_A(t)$	0.1	0.2	0.3	0.2	0.1	0.05	0.03	0.02
$p_B(t)$	-	0.3	0.5	0.1	0.05	0.03	0.02	-
$p_C(t)$	-	-	0.1	0.5	0.3	0.1	-	-

We note that there are six possible routes from node 1 to the destination, node 9, and that two of these (1-2-5-8-9 and 1-4-5-6-9) are equivalent as static routes. The optimist's route would be 1-2-3-6-9, having a minimum journey time of 4, mean of 13.48 and maximum of 32, the realist's route would be 1-2-5-6-9, having a minimum of 6, mean of 12.88 and maximum of 30, whilst the pessimist's route would be 1-4-7-8-9, having a minimum of 12, mean of 17.6 and maximum of 24. Considering the $F_r(t)$ ($r = 1, \ldots, 6$), we find that for $t \leq 9$, the route 1-2-3-6-9 offers maximum reliability, whilst for $t \geq 10$, it is 1-2-5-6-9 which is best.

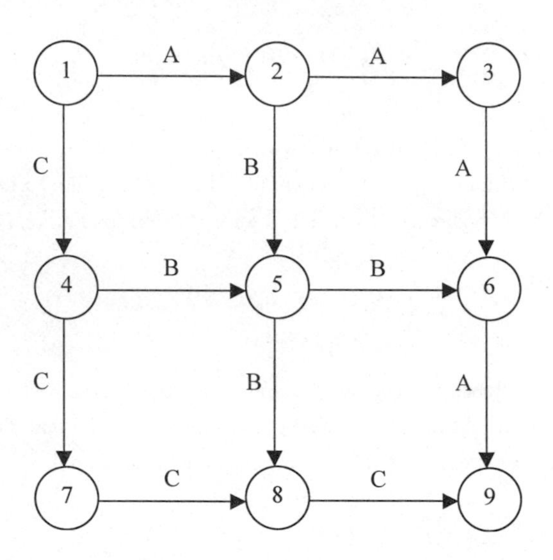

Figure 2 : a test network.

Turning now to the dynamic routeing problem we apply the algorithm described in section 4. It can be seen that it is only at nodes 1, 2, 4 and 5 that there is any choice of exit link. The backward pass starts from node 9, and can proceed in a number of ways, one of which is to treat the nodes in the order: 9, 8, 7, 6, 3, 5, 4, 2, 1. The results of the application of equation (2), to determine the $F_k{}^*(t)$ and optimal successor nodes $L_k{}^*(t)$ are shown in Table 2.

Thus, if a driver starting from node 1 had a target time of $t = 11$ to reach node 9, we can see from Table 2 that he should go initially to node 2 (since $L_1{}^*(11) = 2$). If it turns out that his travel time along link (1, 2) is 4 then he should go next to node 5 (as $L_2{}^*(7) = 5$) whereas if his travel time is 5, then he should go to node 3 (as $L_2{}^*(6) = 3$). The initial reliability is $F_1{}^*(11) = 0.3336$ which we note is marginally greater than the maximum reliability of 0.3303 offered by static routeing (the optimal static route for $t = 11$ being 1-2-5-6-9). This is due to having the flexibility of deciding on the remainder of the route at node 2, depending on the time taken to reach node 2. The advantage of dynamic routeing over static routeing is very small here, but may be more substantial in larger networks.

Table 2 : values of $F_k*(t)$ and (for k = 1, 2, 4 and 5) $L_k*(t)$, for dynamic routeing and, in the last column, $\max_r F_r(t)$ from static routeing.

t	$Fk*(t)$ and (in brackets) $Lk*(t)$									$\max_r F_r(t)$
	k=9	k=8	k=7	k=6	k=3	k=5	k=4	k=2	k=1	
1	1	0	0	.1	0	0	0	0	0	0
2	1	0	0	.3	.01	0	0	0	0	0
3	1	.1	0	.6	.05	.03 (6)	0	.001 (3)	0	0
4	1	.6	0	.8	.15	.14 (6)	0	.007 (3)	.0001 (2)	.0001
5	1	.9	0	.9	.31	.34 (6)	.009 (5)	.028 (3)	.0009 (2)	.0009
6	1	1	.01	.95	.5	.575 (6)	.057 (5)	.078 (3)	.0045 (2)	.0045
7	1	1	.11	.98	.67	.748 (6)	.175 (5)	.175 (5)	.0157 (2)	.0157
8	1	1	.42	1	.796	.856 (6)	.358 (5)	.358 (5)	.0430 (2)	.0423
9	1	1	.74	1	.882	.923 (6)	.5538 (5)	.5538 (5)	.1006 (2)	.0929
10	1	1	.93	1	.938	.966 (6)	.7101 (5)	.7101 (5)	.1983 (2)	.1918
11	1	1	.99	1	.972	.989 (8)	.8215 (5)	.8215 (5)	.3336 (2)	.3303
12	1	1	1	1	.9885	.998 (8)	.8984 (5)	.8984 (5)	.4843 (2)	.4825
13	1	1	1	1	.9955	1 (8)	.9487 (5)	.9487 (5)	.6254 (2)	.6244
14	1	1	1	1	.9984	1 (8)	.9773 (5)	.9773 (5)	.7431 (2)	.7427
15	1	1	1	1	.9996	1 (8)	.9910 (5)	.9910 (5)	.8342 (2)	.8341
16	1	1	1	1	1	1 (8)	.9967 (5)	.9967 (5)	.9003 (2)	.8999
17	1	1	1	1	1	1 (8)	.9990 (7)	.9989 (5)	.9441 (2)	.9433
18	1	1	1	1	1	1 (8)	1 (7)	.9997 (5)	.9706 (2)	.9696
19	1	1	1	1	1	1 (8)	1 (7)	1 (5)	.9855 (2)	.9846
20	1	1	1	1	1	1 (8)	1 (7)	1 (5)	.9933 (2)	.9927
21	1	1	1	1	1	1 (8)	1 (7)	1 (5)	.9976 (4)	.9968
22	1	1	1	1	1	1 (8)	1 (7)	1 (5)	.9994 (4)	.9987
23	1	1	1	1	1	1 (8)	1 (7)	1 (5)	.9999 (4)	.9995
24	1	1	1	1	1	1 (8)	1 (7)	1 (5)	1 (4)	.9998

Table 3 : values of the expected link volumes from dynamic routeing.

Link	Volume	Link	Volume
(1, 2)	940.0	(4, 7)	46.0
(1, 4)	60.0	(5, 6)	549.7
(2, 3)	183.5	(5, 8)	220.8
(2, 5)	756.5	(6, 9)	733.2
(3, 6)	183.5	(7, 8)	46.0
(4, 5)	14.0	(8, 9)	266.8

Turning now to the traffic volumes to be expected on each link, we need to assume a demand profile. For the purpose of illustration, we shall assume that the only origin is at node 1 and that the number of drivers $Q_1(t)$ who have a target time of t to arrive at node 9 is given by a symmetric triangular distribution $Q_1(t) = 100 - 10\ |t - 14|$ for $4 \leq t \leq 24$, and zero elsewhere. Then, the recursive application of equation (3) to the nodes in the reverse order to that used in the application of (2), gives the $q_k(t)$ for all nodes. Summing these over t gives the total link volumes shown in Table 3.

6. EXTENSIONS TO THE DEFINITION OF RELIABILITY

The route choice criterion used so far is concerned only with the maximisation of the probability of arriving within the specified time and this is, in some circumstances, unrealistic. For example, suppose that there was a choice between two routes and that, for some particular target time t, $F_1(t) = 0.9999$ whereas $F_2(t) = 0.9998$, but that the expected journey times were respectively 15 and 12 mins. It is unlikely that, in practice, the driver would place so much weight on the maximisation of $F_r(t)$ that he would select route 1, and ignore the appreciable difference in expected journey time.

Furthermore, the objective function is a purely "black and white" one: the utility of a late arrival is zero (however late), and that of an early arrival is unity. It might be more realistic to allow for a more general form of utility function $U_n{}^*(t)$ to take the place of the 0-1 function $F_n{}^*(t)$ used so far.

The use of this more general utility function does not alter the dynamic routeing algorithm in any way. $U_k{}^*(t)$ is now defined to be the maximum expected utility to be obtained at the destination n, given that the driver is currently at node k, with time t remaining. The basic system of equations in (2) becomes:

$$U_k{}^*(t) = \max_l \sum_s p_{kl}(s) U_l{}^*(t-s), \ \forall k, t \qquad (5)$$

with the initial utility vector $U_n{}^*(t)$ being input. Examples of possible forms of $U_n{}^*(t)$ are shown in Figure 3. The form in which $U_n{}^*(t) = -t$ or $-t^2$ for $t < 0$ and zero elsewhere (Figures 3a and 3b) gives an increasing penalty as the amount of lateness increases. When $U_n{}^*(t)$ is a linear function of t (Figure 3c) the problem reduces to the conventional one of the minimisation of expected journey time. This can be shown by induction.

Assume that for any k in (5), all the relevant $U_l{}^*(t)$ in the maximisation can be expressed in the form $\alpha_l + \beta t$. Then $U_k{}^*(t) = \max_l \sum_s p_{kl}(s)\,(\alpha_l + \beta(t - s)) = \alpha_k + \beta t$ where $\alpha_k = \max_l (\alpha_l - \beta\mu_{kl})$ and μ_{kl} is the mean travel time on link (k, l). Hence $U_k{}^*(t)$ is of the same form as the $U_l{}^*(t)$, since α_k does not depend upon t. Therefore, if the utility $U_n{}^*(t)$ for the destination node n is linear in t, the utilities for all other nodes will be too. Therefore, for each k, the optimal successor node does not depend on t, the time remaining, and is chosen purely on the basis of average link travel times.

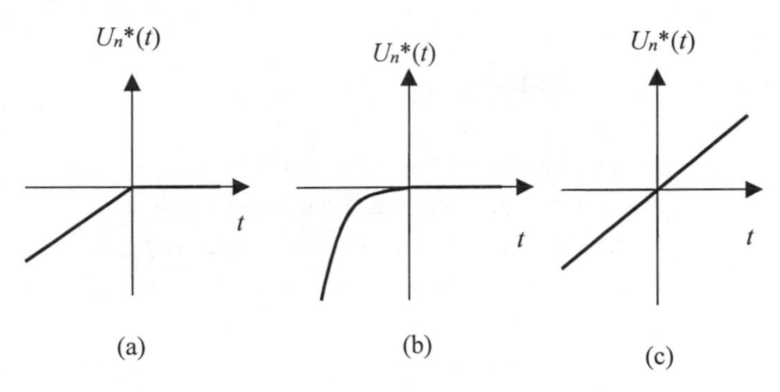

Figure 3 : examples of possible forms of $U_n{}^*(t)$.

We return now to the question of loops in the network. The success of the dynamic programming algorithm in (2) relies on $F_l{}^*(t)$ for all the successor nodes l for node k being known. It is easy to see that in a network containing loops, this will not always be the case. Consider, for example, the simple network in Figure 4.

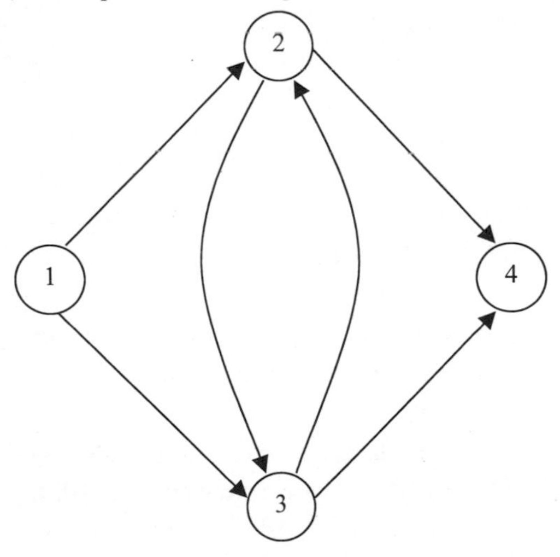

Figure 4 : a network with a loop.

After the specification of $F_4^*(t)$, no further nodes can be dealt with since the determination of $F_2^*(t)$ requires a knowledge of $F_3^*(t)$ and *vice versa*. We suggest that this difficulty can be overcome by the use of an iterative process in which successive estimates $F_k^{(m)*}(t)$ are calculated (m = 1, 2, ...). In any iteration, the maximisation in (2) is carried out using the estimates of the required $F_l^*(t)$ from the current iteration where these are known and from the previous iteration when necessary. Further work is required, however, to investigate the robustness of this proposed scheme.

7. Summary and Conclusions

This chapter has considered the problem of drivers' route choice though a road network when there is known day-to-day variability in link travel times and where the driver's aim is to maximise his probability of arrival at the destination within a specified time. In order to avoid the need for path enumeration, a new algorithm has been developed which assumes a dynamic approach to route choice, whereby the driver need not choose his route at the start of his journey but chooses an exit link at each node, taking account of the time remaining. This algorithm is link-based and is formulated as a many-to-one, Markov process, dynamic programming problem. The outputs from the algorithm consist of the optimal successor node $L_k^*(t)$ and maximised probability $F_k^*(t)$ from all nodes k and for all values of the remaining times t. Furthermore, if a demand profile $Q_m(t)$ is provided, the expected link volumes can be found. The application of the algorithm has been demonstrated on a simple test network.

The simplicity of the algorithm rests on two assumptions: first, the independence of link travel times distributions, so that the process is Markovian; and second, that the network contains no loops. When there are loops, the equation system (2) cannot be solved in a one pass manner and an iterative scheme is required.

Acknowledging the restricted realism of the maximisation of the probability of arrival within a specified time as a route choice criterion, it was shown that a general form of utility $U_n^*(t)$ could be adopted in place of the simple 0-1 form of $F_n^*(t)$ without any alteration to the algorithm. When $U_n^*(t)$ is linear in t, it was shown that the dynamic routeing principle reverts to conventional static routeing based on mean link travel times.

Perhaps the principal use of the method, however, is to provide a systematic and consistent means of carrying out an appraisal of a highway construction or highway improvement scheme in terms of the improvements in travel time reliability. For any O-D pair, the plots of $F^*(t)$ with and without the scheme can be compared, without the need for path enumeration.

REFERENCES

Bell M.G.H. and Y. Iida (1997) *Transportation Network Analysis*. Wiley, Chichester.

Bellman R. (1957) *Dynamic Programming*. Princeton University Press, New Jersey.

Dijkstra E.W. (1959) Note on two problems in connection with graphs. *Numer. Math.* **1**, 269-271.

Howard R.A. (1960) *Dynamic Programming and Markov Processes.* Wiley, Chichester.

CHAPTER 14

A ROUTE RECOGNITION MODEL FOR TRANSPORT NETWORK RELIABILITY ANALYSIS

Yasuo ASAKURA and Eiji HATO

Transoprt Studies Unit
Department of Civil & Environmental Engineering
Ehime University, Matsuyama, 790-8577, Japan
E-mail.asakura@en1.ehime-u.ac.jp

1. INTRODUCTION

Reliability studies in transport network systems are categorized into two approaches. One is to evaluate the connectivity of a pair of nodes in a network. The reliability is defined as the probability of whether an origin and destination (OD) pair is connected for a given set of link connectivities. As shown in Bell and Iida (1998), the connectivity analysis focuses on the topological aspects of a network. Flows in the network are not considered explicitly in this approach.

The other is to evaluate the performance of a network under degraded conditions. The performance measures are, for example, the OD travel time, the OD flow rate, and the total travel times of the entire network. The reliability is defined as the probability of whether a performance measure is sustained within an acceptable level. This approach involves the travel time reliability analysis by Asakura and Kashiwadani (1991), and the degradable transportation systems design by Nicholson and Du (1997). The network reliability studies using performance measures are called performability analyses. The connectivity of an OD pair can be interpreted as the probability of whether travel is possible between the OD pair. Thus, the connectivity approach could be included in the performability approach.

In order to evaluate performance measures, it is necessary to describe flows in a network. This means that network flow models should be included in performability analysis. The previous studies used a network assignment model as the flow model in a degraded network. For example, Nicholson and Du (1997) used the User Equilibrium model with variable demand. Asakura et al. (1999) applied the Stochastic User Equilibrium principle with variable demand to consider probabilistic user's behaviour in a degraded network. Asakura (1999) applied the multi-class Stochastic User Equilibrium model for different driver groups. The User Equilibrium flow models are useful for describing flows in a normal network. It has not been discussed, however, whether the UE concept is still valid in a damaged network.

Although it is necessary to develop an appropriate network flow model considering travel behaviour in a degraded transport system, very few studies can be found on travel behaviour analysis in degraded networks. Juster et al. (1994) reported the user's response to SmarTraveler system in Boston, the cellular phone based travel information service. They found that the ordinary users called during bad weather conditions. This was 3 times more than in normal conditions and 6 times more at the maximum. According to their report, 29% of users changed travel behaviour, which included a departure time change (14.2%), a route change (11.9%), a trip cancel (2.2%) and a time/route change (1.1%), respectively.

Khattak and De Palma (1997) studied the impact of adverse weather conditions on the propensity to change travel decisions using survey data in Brussels. They observed that 54% of the auto users changed their travel choice in response to weather conditions. The influence of weather on mode change was 27%, 61% on departure time change and 36% on route change, respectively. They also applied the ordered probit model of mode/departure time change in adverse weather. Asakura et al. (1997) studied travel choice behaviour in a road network degraded by natural disasters in the Shikoku area of Japan. They found that the 40% of drivers experienced traffic closure due to natural disaster. Of these, 14% cancelled their trip and 86% changed their route. 75% of those who changed route diverted to routes with a lower traffic function. This implies that the drivers are obliged to use the route with higher risk, since the maintenance quality of a lower class of road is relatively lower than a higher class of roads.

Travel behaviour in a degraded network will depend on the user's attributes. In a degraded condition, a driver's knowledge of the network effects largely his/her route choice behaviour. Thus, the objective of this chapter is to study a route choice model under degraded conditions. In particular, we focus on the user's uncertain recognition of network structure and route choice in the recognized network. The difference between an informed driver and a non-informed driver is described comparatively in a route recognition model. The results will be incorporated in a network flow model with different user groups.

2. Route Recognition Model in a Degraded Network

In the deterministic User Equilibrium traffic assignment, a driver is assumed to have perfect knowledge on the conditions of a network. He/she can choose the best route among all possible routes between an OD pair. The set of alternative routes is not explicitly handled in the link flow based algorithm for solving the UE assignment. In the path flow based algorithm such as the simplicial decomposition method, the set of routes is explicitly considered. For example, the column generation phase of the simplicial decomposition algorithm examines the set of possible routes. The candidate route should be added to the set of routes if the travel cost along the route is lower than those of the routes in the existing set. However, a driver's recognition of the set of routes is not considered. All of the possible routes between an OD pair could be viewed as the alternatives.

On the other hand, the set of alternative routes is restricted in Dial's algorithm to efficient paths. An alternative route between an OD pair is called an effective path, on which flows would be loaded. The concept of the efficient path, however, was generated through the calculation process. It might not be directly related to the assumption of a driver's route recognition and choice behaviour. A driver's recognition of the network is not considered in Dial's stochastic network loading algorithm. This is the same for other stochastic network loading methods like the Markov chain model.

Even in normal condition without any failure, a driver may not have perfect knowledge of the configuration of the network. He/she may believe that a link does not exist while it is actually connected. In a degraded condition, some links will be closed to traffic. The network recognition of a driver will become ambiguous in those degraded conditions. In this chapter, we will distinguish the level of network recognition between the normal and the degraded conditions. The level of recognition also depends on whether a driver is informed or not. The difference of those cases are discussed in the following section.

2.1 Network State

The condition of the network with L links is denoted by a network state vector $x = \{x_1, \ldots, x_a, \ldots, x_L\}$. A damaged link is assumed perfectly closed to traffic. The element of the state vector of a damaged link is written as $x_a = 0$. If a link is operated, the element of the state vector of the link is represented by $x_a = 1$. Various states will occur in a network and a network state is distinguished using subscript j as $\boldsymbol{x}_j$. The set of state vector is written as $X = \{\boldsymbol{x}_j\}$. We define the occurrence probability of a state $\boldsymbol{x}_j$ as $p(\boldsymbol{x}_j)$. The range of the probability is $0 \le p(\boldsymbol{x}_j) \le 1$, and the probability satisfies $\sum_{\boldsymbol{x}_j \in X} p(\boldsymbol{x}_j) = 1$. The normal network is represented by $\boldsymbol{x}_0$. Figure 1 depicts examples of the network conditions and corresponding state vectors.

A driver may not always recognize the actual structure of a network. Under degraded conditions, uncertainty of driver's recognition increases and a wrong recognition may often occur. Thus, we distinguish the actual state and the recognized state of a network. Both states are represented by the state vectors $\boldsymbol{x}_j$ and $\boldsymbol{x}_j^*$, respectively. Figure 2 shows the examples of the actual and the recognized states of a network. Usually, those two states are different. The recognized state will be consistent with the actual state if a driver has sufficient knowledge and information.

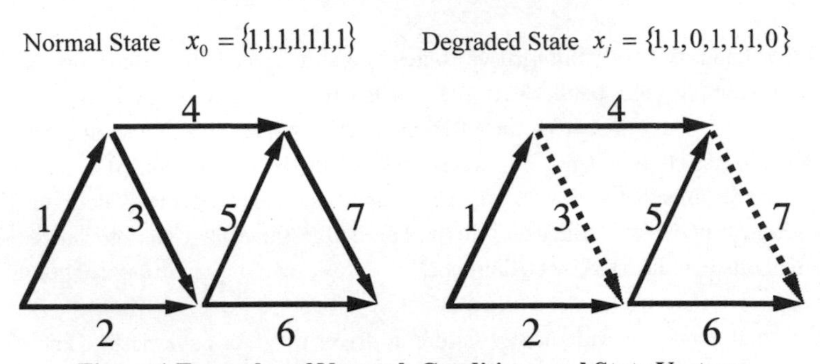

Figure 1 Examples of Network Conditions and State Vectors.

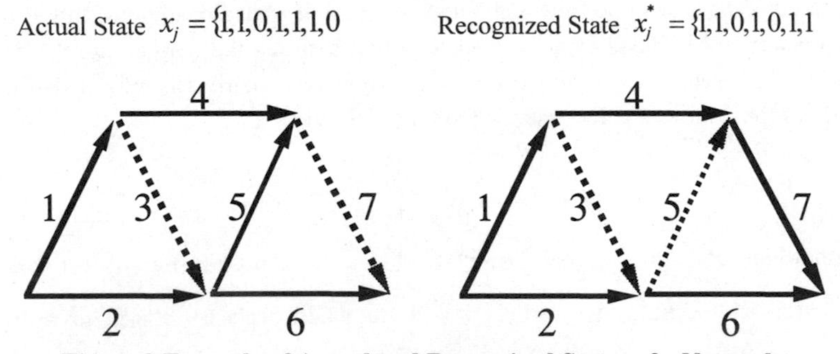

Figure 2 Example of Actual and Recognized States of a Network.

2.2 Assumption for Driver's Network Recognition and Route Choice

We denote the set of routes (the path set) as $K_{rs}(\boldsymbol{x}_j)$ between origin r and destination s in the actual state $\boldsymbol{x}_j$ of a network, and the set of route in the recognized state as $K_{rs}(\boldsymbol{x}_j^*)$. For the network shown in Figure 2, there exists only one path in the actual network. On the other hand, two paths can be found in the recognized network.

(1) Normal Network
In normal conditions without any deteriorated links, we assume that all drivers have perfect knowledge of the network and they can recognize the network correctly. This means that the recognized state x_0^* is equal to the actual state x_0 in the normal network. Thus, the set of the recognized routes in the normal network $K_{rs}(x_0^*)$ is equivalent with the set of actual routes in the normal network $K_{rs}(x_0)$ for any OD pairs. All drivers can choose the route with minimum travel cost from the recognized set $K_{rs}(x_0^*)$, which is also equal to the minimum cost route in the actual network.

(2) Degraded Networks
In the network with some deteriorated links, a driver's recognition of the network becomes uncertain even if he/she has sufficient experience in the normal network. The recognized state of the network becomes different from the actual state of the network. However, such inconsistency will not occur if the actual condition of the network is sufficiently known to a driver. We distinguish two types of drivers; one who is an informed driver and the other who is a non-informed driver.

An informed driver is assumed to be able to recognize the actual network condition correctly if some of the links in the network are deteriorated. The recognized state x_j^* is equal to the actual state x_j as we have assumed for the normal network. Thus, the set of recognized route $K_{rs}(x_j^*)$ at state x_j^* is equal to the set of actual routes $K_{rs}(x_j)$ at the state x_j. An informed driver can choose the route with minimum travel cost from the recognized path set $K_{rs}(x_j^*)$. When there is no available route in the path set, the informed driver will stop travelling.

A non-informed driver who is assumed not to be able to recognize the actual network correctly may lose his/her way in a deteriorated network. The recognized network state of a non-informed driver may be largely different from the actual network state. Thus, the set of the recognized route of a non-informed driver is not always equal to the set of the actual route. A route in the recognized set may not exist in the actual network.

A non-informed driver will also choose the route with the minimum travel cost from the recognized path set $K_{rs}(x_j^*)$. This is the same route choice behaviour as an informed driver. The difference is, however, that a non-informed driver will travel as if he/she is in a maze because the chosen route in a recognized set may not be physically connected in the actual network.

Here, we consider the route choice behaviour of a non-informed driver where the chosen route from the recognized set does not exist in the actual network. A non-informed driver continues travelling along the chosen route while the link in the chosen route is in operation. When a non-informed driver reaches the end of the existing route where the next link on the route is not connected, he/she tries to find a route from the current position in the network. The new route with the minimum travel cost must be found in the recognized path set. If at least one route in the recognized network is connected between the OD pair, a non-informed driver can reach his/her destination after en-route switching several times. When all routes in the recognized set do not actually exist, a non-informed driver has to stop travelling after some wasted trials.

Let us explain the route choice behaviour of a non-informed driver using an example network in Figure 3. The numbers in the figure are the link travel costs. A non-informed driver starts from the origin r and goes to the destination s. Along the minimum travel cost route, the driver uses the link r-c. The successive link c-s is, however, not an existing link. Then, the driver has to find his/her route at the node c to the destination s in the recognized network without link c-s. The available route from node c to destination s is the one which returns back to the origin r and uses links r-a, a-b and b-s.

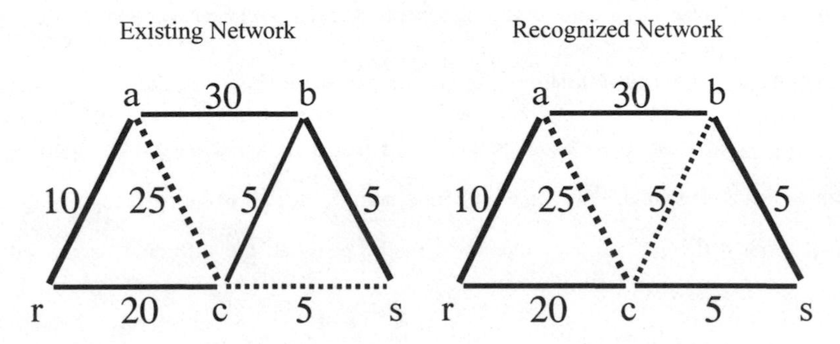

Figure 3 Route Choice Behaviour of a Non-Informed Driver.

3. Evaluation of Network Performance

The driver's route choice model presented in the previous section is involved in the evaluation process of network performance. We denote a network performance measure $Z(\boldsymbol{x}_j)$ for a network state $\boldsymbol{x}_j$. For example, the total travel cost in a network is taken as a performance measure. The performance measure is also defined as $Z_{rs}(\boldsymbol{x}_j)$ for an OD pair r-s in the network. The connectivity measure is interpreted as a performance measure between an OD pair, which means that $Z_{rs}(\boldsymbol{x}_j)=1$ if the OD pair is connected or $Z_{rs}(\boldsymbol{x}_j)=0$ if it is disconnected. When we evaluate the travel time reliability between an OD pair, the performance measure $Z_{rs}(\boldsymbol{x}_j)$ can be defined as follows; $Z_{rs}(\boldsymbol{x}_j)=1$ if the journey is possible within an acceptable travel time, or $Z_{rs}(\boldsymbol{x}_j)=0$ if it is impossible. The OD travel time itself can be used as a performance measure.

As mentioned above, the state of a network is uncertain. The occurrence of a network state $\boldsymbol{x}_j$ is represented by the state probability $p(\boldsymbol{x}_j)$. Thus, the expected value of a performance measure is used for evaluating network performance under degraded conditions. The expected value is defined as,

$$E[Z]=\sum_{\boldsymbol{x}_j \in X} p(\boldsymbol{x}_j)Z(\boldsymbol{x}_j). \tag{1}$$

If the number of the state vectors is limited, it is not so difficult to enumerate all possible state vectors. The performance measure is calculated for each network state and the expected value $E[Z]$ can be directly evaluated using the equation (1).

The enumeration of all possible state vectors, however, requires huge computational cost for a large scale transport network. When each link in a network is randomly degraded and the occurrence of failure of a link is independent, the number of the possible state vectors of a network with L links becomes 2^L. This makes it impossible to enumerate all possible state vectors for evaluating the expected value of a performance measure.

Without enumerating all possible state vectors, we can approximate the expected value of a performance function. A typical approximation method is the application of the Monte-Carlo simulation. The expected value of a performance measure can be approximated using randomly generated network state vectors. Instead of the random generation of state vectors, Li and Silvester (1984) proposed to approximate the upper and the lower bounds of the expected value using the J most probable state vectors. The upper and the lower bounds are approximated in the following equations.

$$Z^{upper} = \sum_{j=0}^{J} Z(\boldsymbol{x}_j)p(\boldsymbol{x}_j) + (1 - \sum_{j=0}^{J} p(\boldsymbol{x}_j))Z(\boldsymbol{x}_0) \tag{2}$$

$$Z^{lower} = \sum_{j=0}^{J} Z(\boldsymbol{x}_j)p(\boldsymbol{x}_j) + (1 - \sum_{j=0}^{J} p(\boldsymbol{x}_j))Z(\boldsymbol{x}_W) \tag{3}$$

The state probability is arranged in the order of the degree of the occurrence probability as:

$$p(\boldsymbol{x}_0) \geq p(\boldsymbol{x}_1) \geq p(\boldsymbol{x}_2) \geq \ldots\ldots\ldots\ldots p(\boldsymbol{x}_J) \geq \ldots\ldots \tag{4}$$

$Z(\boldsymbol{x}_0)$ and $Z(\boldsymbol{x}_W)$ denote the best and the worst values of the performance measure. For example, $Z(\boldsymbol{x}_0) = 1$ and $Z(\boldsymbol{x}_W) = 0$ for evaluating connectivity. When we use the performance measure like travel cost, which becomes larger for worse network states, we should be careful that the upper and the lower bounds are replaced by each other. This is because the value of $Z(\boldsymbol{x}_W)$ is larger than that of $Z(\boldsymbol{x}_0)$, and Z^{lower} is consequently greater than Z^{upper}.

For all cases, the second term of the above equations (2) and (3) becomes smaller when we calculate larger numbers of the state vectors. This means the gap between the upper and the lower bounds becomes narrower for larger numbers of iterations. The approximation procedure can be summarized as follows.

<Step.1> Set iteration counter J=0.

<Step.2> Take the J-th most probable state vector $\boldsymbol{x}_J$.

<Step.3> Calculate the performance measure $Z(\boldsymbol{x}_J)$ for the state vector $\boldsymbol{x}_J$.

<Step.4> Evaluate the upper and the lower bounds of the expected values using equations (2) and (3).

<Step.5> Check the convergence. If the difference between the upper and the lower bounds is small, the expected value of the performance measure is calculated as:

$$Z_{approx.} = (Z^{upper} + Z^{lower}) / 2 .$$

Otherwise, J=J+1 and return back <Step.2>.

4. NUMERICAL EXAMPLE

Using a small size network, we will show a numerical example of evaluating network performance under degraded conditions. A test network has 9 nodes and 12 directed links. An origin node and a destination node are denoted by r and s, respectively. Links are separated into two types. A solid line in a network represents a link that is not degraded. A dashed line means a link that is possibly deteriorated. At least one route remains in a network even if all dashed lines are not connected. All links have sufficient capacity and congestion is not generated.

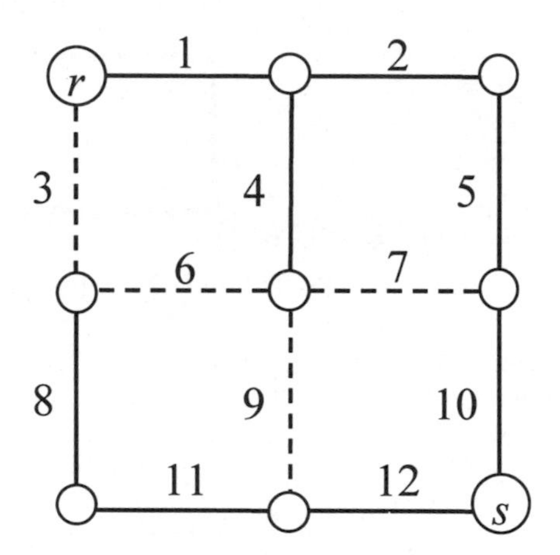

Figure 4 Network Configuration.

The probability of whether a link is sustained at a normal condition is denoted by p_a. $p_a = 1$ for solid links and $0 < p_a \leq 1$ for dashed links. We assume that failure in each link happens independently. Thus, the state probability $p(\boldsymbol{x}_j)$ is written as:

$$p(\boldsymbol{x}_j) = \prod_a p_a^{x_a}(1 - p_a)^{1-x_a}. \tag{5}$$

Drivers are separated into two groups: informed drivers and non-informed drivers. In addition to the behavioural assumptions described in the previous sections, we also assume that a non-informed driver always recognizes the network as it is under normal conditions. This implies the recognized path set of a degraded network $K_{rs}(\boldsymbol{x}_j^*)$ is equal to the actual path set of the normal network $K_{rs}(\boldsymbol{x}_0)$. The shortest path choice behaviour is assumed for both types of drivers.

A route choice example is shown in Figure 5. An informed driver could recognize the state of the network correctly and use the proper route. A non-informed driver would recognize the network as if the degraded links were in operation. His chosen route would not actually exist, and he would be obliged to drive a wasteful route.

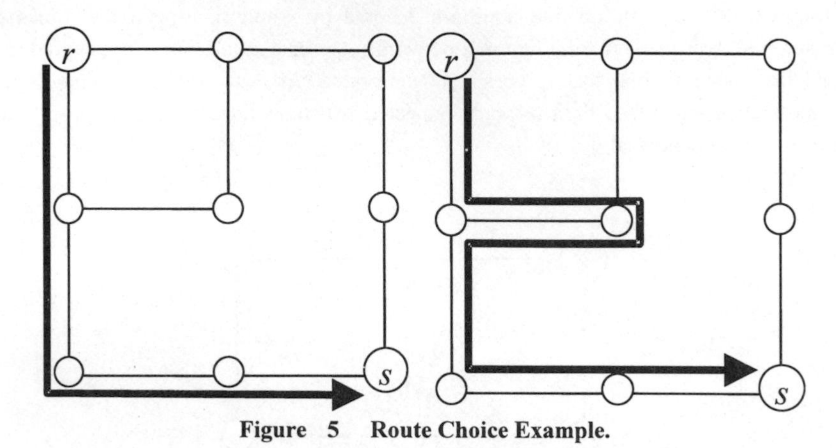

Figure 5 Route Choice Example.

We examined network performance using the OD travel cost as a performance measure. The expected value of the OD travel time for an informed driver and that for a non-informed driver can be calculated from the following equations, respectively.

$$E[t_1] = \sum_{x_j \in X} p(x_j) t_1(x_j), \qquad E[t_2] = \sum_{x_j \in X} p(x_j) t_2(x_j) \tag{6}$$

Here, $t_1(x_j)$ and $t_2(x_j)$ are the OD travel cost at network state x_j for an informed driver and a non-informed driver, respectively.

Figure 6 shows the expected travel costs $E[t_1]$ and $E[t_2]$ for different levels of network deterioration. The value of link connectivity p_a of a dashed link is used as a parameter. When a network is physically robust and the link connectivity is almost equal to 1.0, the difference between $E[t_1]$ and $E[t_2]$ is small. The expected travel costs increase as the link connectivity decreases. The increase rate of the expected travel cost for a non-informed driver is larger than that for an informed driver. If the link connectivity reduces to 0.1, the expected travel cost for a non-informed driver grows exponentially and becomes double that of the normal network state.

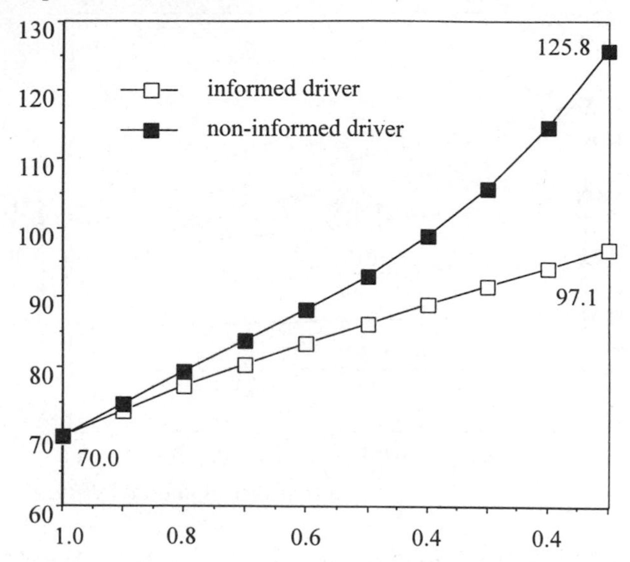

Figure 6 Comparison of Expected Travel Cost.

We can also discuss the effects of providing information under degraded network conditions. We assume that the total number of drivers is fixed and the percentage of informed drivers is a parametric variable. That is,

$$\omega q_1 + (1-\omega) q_2 = const. \tag{7}$$

where q_1 and q_2 denote the number of informed and non-informed drivers. ω denotes the percentage of informed drivers, which varies the range of $0 \le \omega \le 1$. The expected total cost for all drivers is,

$$E[t] = \omega E[t_1] + (1-\omega) E[t_2] = E[t_2] - \omega(E[t_2] - E[t_1]). \tag{8}$$

This is a linear function of $E[t_1]$ and $E[t_2]$. Figure 7 shows the effects of providing information to drivers in terms of reducing travel cost. For different combinations of the link connectivity and the percentage of informed drivers, the expected total travel cost is depicted. Due to the large difference between $E[t_1]$ and $E[t_2]$, it is effective to provide information for a network when link connectivity is small. Providing information to drivers will remove uncertainty of travel, and it will contribute to reduce the expected travel cost.

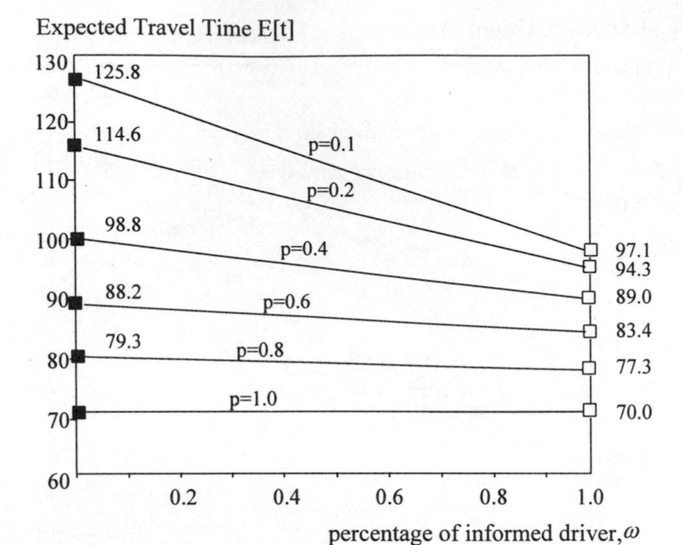

Figure 7 Effects of Providing Information for Reducing Travel Cost.

5. Conclusion and Further Studies

In this chapter, we have studied a driver's behavioural model in a deteriorated network. Focusing on the difference between the recognized network and the actual network, the route choice behaviour of two types of drivers was comparatively discussed. For a non-informed driver, the gap between the recognized path set and the actual path set causes wasteful driving in the network. If a network is seriously deteriorated, such a gap becomes larger and a non-informed driver will be obliged to drive as in a maze. However, the model shown in this chapter remains at a primitive level. There are many topics which should be studied in the future.

A driver's recognition level of a degraded network state depends on the degree of deterioration. Even if a driver is familiar with the network at a normal condition, it will be difficult for him/her to recognize a network state correctly when the network is seriously damaged and many links are closed to traffic. As a result, the percentage of those who can recognize the network correctly will decrease if the network is significantly damaged. The percentage of informed drivers might be handled as an endogenous variable in relation to the degree of link connectivity, which represents the level of deterioration.

In addition to the recognition problem, route choice behaviour in a deteriorated network should be studied from another direction. A driver may eliminate his/her possible routes in the network. The elimination by aspects (EBA) model can be applied to describe this situation. The network EBA studied by D'Este (1997) is different from the original EBA by Tversky (1972). An example of EBA is stated as follows. A driver first eliminates deteriorated links (or links recognized as being deteriorated). Then, he/she eliminates the routes using the most important attribute. Long distance routes may be eliminated from the path set. The acceptable level depends on the degree of deterioration. A diversion route with longer distance may be accepted if the network is significantly damaged. When only one path remains in the path set, the driver chooses it. Otherwise, he/she repeats the elimination for the next important attribute. This procedure of route choice can be combined with the route recognition model.

We have found that providing information in a deteriorated network would be effective in terms of reducing travel cost. However, congestion is not considered in the model. The effects of providing information in an actual deteriorated network might not be so large because of the congestion due to diverted traffic. The flow dependent analysis should be further studied in the future.

Acknowledgement

The authors would like to show their appreciation to Mr. Daisuke Ochi, a master course student of Ehime University, for his works in numerical examples.

References

Asakura, Y., Kahiwadani, M., Takagi, K. and Fujiwara, K.(1997) Empirical Studies on Travel Choice Behaviour in a Road Network Degraded by Natural Disasters. *Infrastructure Planning & Management*, No.14, pp.371-379.

Asakura, Y. (1999) Evaluation of Network Reliability using Stochastic User Equilibrium. *Journal of Advanced Transportation* (in press)

Asakura, Y., Kashiwadani, M., Hato, E. (1999) Flow Model and Performability of a Road Network under Degraded Conditions. *Proceedings of the 14th International Symposium on Transportation and Traffic Theory*, (Ceder, A. ed.) Pergamon, pp.257-281.

Bell, M.G.H. and Iida, Y. (1997) *Transportation Network Analysis*. John Wiley & Sons.

D'Este, G. (1997) Hybrid Route Choice Procedures in a Transport Network Context. *Journal of the Eastern Asia Society for Transport Studies*, Vol.2, No.3, pp.737-752.

Juster, R.D., Wilson, A.P. and Wensley, J.A. (1994) An Evaluation of the SMARTRAVELER ATIS Operational Test. *Paper presented at the 4th annual meeting of IVHS America*, Atlanta, Georgia, April 1994.

Khattak, A. and De Palma, A. (1997) The Impact of Adverse Weather Conditions on the Propensity to Change Travel Decisions: A Survey of Brussels Commuters. *Transpn. Res.-A*, Vol.31, No.3, pp.181-203.

Li, V.O.K. and Silvester, J.A. (1984) Performance Analysis of Networks with Unreliable Components. *IEEE Trans. on Communications*, Vol.COM-32, No.10, pp.1105-1110.

Nicholson, A. and Du, Z.P. (1997) Degradable Transportation Systems: An Integrated Equilibrium Model. *Transpn. Res-B.*, Vol.31, No.3, pp.209-223.

Tversky, A. (1972) Elimination by Aspects: a theory of choice. *Psychological Review*, No.79, pp.281-299.

CHAPTER 15

USING NETWORK RELIABILITY CONCEPTS FOR TRAFFIC CALMING: PERMEABILITY, APPROACHABILITY AND TORTUOSITY IN NETWORK DESIGN

Michael A P Taylor

Transport Systems Centre, University of South Australia

Adelaide, South Australia

INTRODUCTION

A recent pair of papers (Taylor, 1999ab) discussed the use of dense network traffic models in network reliability studies and introduced the concept of PAT (Permeability, Approachability and Tortuosity) analysis for traffic impact studies and the appraisal of network layouts and traffic management plans. This chapter carries the discussion a stage further, through the application of the PAT analysis to a real network with known traffic and environmental problems. This provides a useful indication for the use of the PAT indices comparing different traffic management plans for a dense network of local streets and arterial roads

Evaluation of traffic management plans in an area can require study of the degree of connectivity and extent of reliability of the network and likely effects on driver route choice behaviour and trip timing. As indicated throughout this book, network reliability may generally be seen from three perspectives: connectivity, travel time reliability and capacity reliability. The connectivity of a network indicates the ability to move efficiently between any two points. For local area traffic studies connectivity may relate to the relative performance of different traffic calming schemes, such as speed control devices and street systems and turn bans at intersections, to (say) restrict through traffic movement in some parts of the network. The other useful perspective for local area studies is time reliability, which is concerned with the perception and predictability of individual journey times through the network, which may vary over time of day in response to changes in traffic conditions and congestion. Time reliability will also be affected by factors such as the level and spread of traffic congestion, traffic control systems, parking provisions, interactions with public transport services, origin-destination patterns, and the presence of bottlenecks.

TRAFFIC CALMING

Contemporary traffic planning practice focuses on assessment of the traffic, economic, environmental, energy and social impacts of new land use developments and traffic management and control schemes, with requirements for community acceptance that can only be achieved through active programs of community participation in the planning process (e.g. Westerman, 1990). Traffic planning in the 1970s and 1980s was driven by the concept of a 'road hierarchy', with major routes (freeways and arterial roads) providing mobility (vehicle travel across a region) and local streets and roads providing access to dwellings, shops and businesses. This useful concept remains in place but, as indicated by Brindle (1989) and Westerman (1990), there are problems with roads in the middle of the hierarchy (e.g. the collector roads) that have both mobility and access functions. In the early 1990s the concept of 'traffic calming' became significant (Brindle 1991), with the objective of restricting the nature of traffic movements (e.g. speed of travel) to levels deemed commensurate with the surrounding environment and meeting community needs and aspirations whilst still providing for traffic throughput. Traffic congestion also emerged as a significant issue for many cities around the world, with an especial interest in non-recurrent or 'incident-based' congestion (e.g. Pfefer and Raub 1998). Analytical methods for assessing likely deficiencies in network performance under different traffic conditions, including incident situations, would be most helpful for traffic management and control purposes.

Marshall (1998) provides a comprehensive review of issues affecting the design of road networks and street patterns as part of an overall urban design process and within an integrated transport planning framework. He highlights the importance of the road network in the general layout of cities through the creation of 'accessibility fields' which influence the distribution of land uses and activities in an urban area. This includes an emphasis on considerations of 'network permeability' and comparisons of 'tributary-style' residential area networks with classical or neo-classical grid layouts. Marshall argues strongly for the need to take account of road type, pattern and functional classification as separate entities in network design, to ensure flexibility in the design process and the ability to cater for different types of trip and traffic movement (e.g. pedestrian and public transport movement, and through traffic and local traffic). Guidance on network layouts needs to embrace both traffic engineering and urban planning expertise, using clear and unambiguous definitions of network characteristics. The PAT indices discussed in this chapter provide some assistance in meeting these goals.

Further, there is increasing emphasis on 'traffic calming' as a planning objective aimed at mitigating the adverse impacts of motor vehicles in urban areas, including the safety impacts of those vehicles. Traffic calming may be seen as providing a practical link between transport economics and planning and traffic engineering, especially from the road safety perspective. Traffic calming is a composite term representing a number of policy and planning measures, some with physical and environmental manifestations, others with social and cultural connotations. Traffic calming may also be applied at a variety of levels, from the site-specific to the metropolitan level. Brindle (1991) described the 'Darwin matrix' as a framework for relating and classifying traffic calming measures. A version of this matrix is given in Table 1. This is based on three spatial levels for the application of traffic calming measures: local, intermediate and macro - which determine the scope of the measure, and two types of measures: technical measures involving physical or environmental treatments and 'ethical' measures involving social and cultural factors. Brindle's Darwin matrix thus provides an overview of alternative transport policies for road traffic systems.

Traffic calming at the local or regional level requires analytical decision support tools for effective planning and implementation. The appropriate tools are dense network models (Taylor, 1999ab). Dense networks are defined as networks in which all of the street and road sections in the area need to be considered for inclusion. They possess the following characteristics:

Table 1: The 'Darwin matrix' of traffic calming measures (derived from Brindle (1991))

Scope of measure	**Type of measure** Physical/environmental ('technique') **E**	 Social/cultural ('ethos') **C**
L Local (street, neighbourhood)	**LE** Local area traffic management, speed control devices, site-specific speed and accident physical countermeasures	**LC** 'Neighbour Speed Watch' programs, community action, school road safety programs, attitudinal change
I Intermediate (zone, precinct, corridor, regional)	**IE** Environmentally-adapted through roads (e.g. Northern Europe), shared zones, lower speed limit zones, pedestrianised shopping precincts, corridors (e. g. Westerman (1990) type II roads), road pricing zones, parking policies, driver information systems	**IC** Voluntary behaviour change, e.g. for mode choice, speed and time of travel, car pooling, speed limit enforcement zones and corridors
M Macro (city-wide)	**ME** Travel demand management (TDM), transportation systems management (TSM), total system measures (e.g. fares policy, city-wide road pricing, regional speed limit enforcement), changing urban form and structure, parking policies, traveller information systems	**MC** Cultural change, restrictions on travel choices, population decline, futurology, telecommuting, travel blending, media publicity campaigns on effects of speeding and alcohol

(a) there is a general one-to-one correspondence between the links in the network and the actual road and street sections in the study area;

(b) the turning movement flow (i.e. how many vehicles turn right, left or travel straight through an intersection) is the basic measure of traffic volume;

(c) trip generation can take place along the links of the network, rather than at specially designated nodes (the 'zone centroids' of strategic networks). This matches the real world process of trip generation;

(d) turning movement delays at intersections are important and may dominate in the determination of travel times;

(e) traffic management controls and devices need explicit recognition and possible differentiation within the model, and

(f) the dense network model is a hybrid of transport planning, traffic flow theory and traffic engineering.

The basic unit of a dense network is the turning movement. Associated with each turning movement are the movement volume (perhaps by vehicle type), delay, queue length and capacity. Link volumes may be found by summing and comparing the turning movement volumes entering and leaving the link. Travel time on a link consists of the time required to traverse the link itself, and the delay times associated with the turning movements from that link. A dense network model can distinguish between (1) different classes of roads and streets, (2) various design standards for streets, (3) different intersection controls, (4) turning and through traffic at intersections, (5) different vehicle types, and (6) different classes of traveller (e.g. local or through traffic, single occupant or multi-occupant vehicles, etc).

Network Reliability in Traffic Calming Studies

Bell and Iida (1997) described transport network reliability in terms of two main aspects: network connectivity and travel time reliability, and this description has been followed throughout this book (e.g. previous Chapters 8, 9, 10 and 12 as well as this chapter). Connectivity considers the probability that traffic can reach a given destination at all, whilst travel time reliability is seen as the probability that traffic can reach a given destination within a specified time period.

In traffic calming studies, these concepts of network reliability can be applied in terms of individuals' perceptions of travel times (or costs) and differential network connectivity (where the traffic calming plan may seek to restrict access to some parts of the network for some road users). Perceived travel times may be based on a driver's day-to-day experiences of using a given route or specific links. These will range from a complete lack of knowledge on the part of the stranger, who will rely on maps, direction signs and driver information systems (DIS), to the regular, experienced user who may have good knowledge of the connectivity provided on the minor road network and who may use nuances gained from traffic conditions at one place in the network to infer likely downstream conditions, on the basis of past experiences.

Differential network connectivity is a useful concept for the design of local area traffic management schemes, where the objectives may be to provide reasonable access to or from given locations in the area, but to restrict through traffic usage of the area. Local access may be described in terms of 'approachability' (a form of micro-level accessibility to facilities) or 'tortuosity' (the extent of deviation from a direct path) whilst the degree of difficulty of through movement may be described in terms of a network or area permeability.

Sources of Travel Time Variations

Wong and Sussman (1973) considered the sources of variations in individual travel times on a route section, and the consequent implications for errors in travel time observations. They divided travel time variations into three broad types:

(a) *regular condition-dependent variations*, arising from differences in travel conditions between times of day, days of the week, and seasons of the year. These variations may usually be explained in terms of factors such as traffic flow rates, climate, and travel demand patterns, and they usually occur on a network-wide level;

(b) *irregular condition-dependent variations*, which arise from spasmodic, unpredicted changes in travel conditions on a route, such as accidents, breakdowns, signal failures, roadworks and weather conditions. In modern parlance, these factors would be described as incidents, and

(c) *random variations*, which are isolated, unpredicted events affecting only a small number of vehicles, e.g. sudden braking to avoid a pedestrian crossing the road.

The distinction between irregular condition-dependent variations and random variations is largely a matter of the number of vehicles affected. The propensity for a link to exhibit irregular condition-dependent variations and random variations gives an indication of its travel time reliability.

Modelling Travel Time Variations

In a series of observational studies of travel time variations, Richardson and Taylor (1978) and Taylor (1982) built on the work of Herman and Lam (1974) to examine daily variations of travel times on links in different networks. This research suggested that the distribution of individual travel times about the mean travel time (t) was related to the mean value by the equation

$$s = \gamma\sqrt{t} \tag{1}$$

where s is the standard deviation of the travel time variations and γ is a constant, possibly varying between links of different types. Equation (1) indicates that the dispersion of the distribution of individual travel times will increase with increasing value of the mean travel time, according to a square root relation. An assumption behind this equation is that travel times on adjacent links are independent. Extensive surveys and data analysis (Richardson and Taylor (1978), Taylor (1982)) provided statistical evidence strongly supporting this assumption.

Further, the distribution of travel time variations may be described, under different conditions, by either the normal distribution or the log-normal distribution. The normal distribution is more suited to heavily congested links. Less congested links had variations better described by the log-normal distribution (Richardson and Taylor (1978), Taylor (1982)). This result may be explained in terms of the symmetry of the normal distribution. For this distribution, the probabilities of observing travel times much longer or much shorter than the mean are the same. However, short travel times are curtailed by the free flow time, a finite lower bound on the feasible link travel time. In less congested conditions, some individual long travel times are still possible, but the corresponding short travel times may not be. Thus the distribution will be skewed to the upper tail. In congested conditions when the mean is sufficiently greater than the free flow travel time, symmetry can apply, at least for practical purposes.

This result for travel time variations is built into the TrafikPlan model (Taylor, 1992), where it is used in its stochastic user equilibrium (SUE) submodel, either as a probit-based SUE or log-normal-based SUE. The advantage of this approach is that the distribution of individual travel times is functionally defined in terms of the mean travel time (by equation (1)). The coefficient γ may be estimated for a given network, as indicated in Taylor (1979). Both individual perceptions of travel time and the reliability of travel time on links and routes in a TrafikPlan network can be represented by the distribution defined by equation (1).

Measures of Network Permeability

As indicated by Bell and Iida (1997) and Chapters 8, 9, 10, 11, 12 and 14, network reliability can be assessed in terms of connectivity, being the existence of one or more connected paths between a given origin and a destination. The concept of differential connectivity can be applied in local area traffic management, not so much as to examine absolute connectivity (as above), but to provide different levels of connectivity for different traveller groups or road user groups. For instance, a traffic management plan for a residential street network may attempt to remove through traffic from that area whilst retaining a high (or at least acceptable) degree of accessibility to or from trip ends in the area. The traffic management plan may seek to remove the through traffic by using barriers and street closures, etc to make the area impermeable to through traffic, or it may seek to increase travel times on the local streets so that arterial road paths are clearly preferable. Different networks and traffic management plans may provide different levels of permeability, which may be compared to the general concepts of network reliability. In addition, it may also be possible to compare alternative traffic management plans for an area by considering their relative permeabilities.

Traditional network graph theory provides some parameters for describing the physical state of a network (e.g. Eliot Hurst, 1974). By considering the parameters of physical connectivity and adapting them to consider travel times for different journeys through a network, Taylor (1999b) proposed three indices that may be applied to describe the state of a local area network relative to its surrounding arterial road network. The indices are:

1. a permeability index that indicates the relative attractiveness of an area to through traffic;
2. an approachability index that indicates the relative ease of access to or from the local area from or to the surrounding main roads, and
3. a tortuosity index that indicates the difficulty of movement between locations inside the local area.

These indices are defined as follows.

Consider the street system in a region to be a network (*N*) composed of two sub-networks *H* and *G*. The sub-network *H* represents the main roads (say the arterial roads surrounding a traffic cell or a neighbourhood). Sub-network *G* is the network of local streets in the cell. Figure 1 represents such a street system, in this case the TrafikPlan network for the suburb of

Ashburton in Melbourne, Australia. The full network N is the union of the two sub-networks, i.e.

$$N = H + G \qquad (2)$$

Network reliability considerations arise because the travel times in the two sub-networks, and thus the relative attractiveness of alternative routes through the two sub-networks, change in response to traffic conditions and congestion levels on the network.

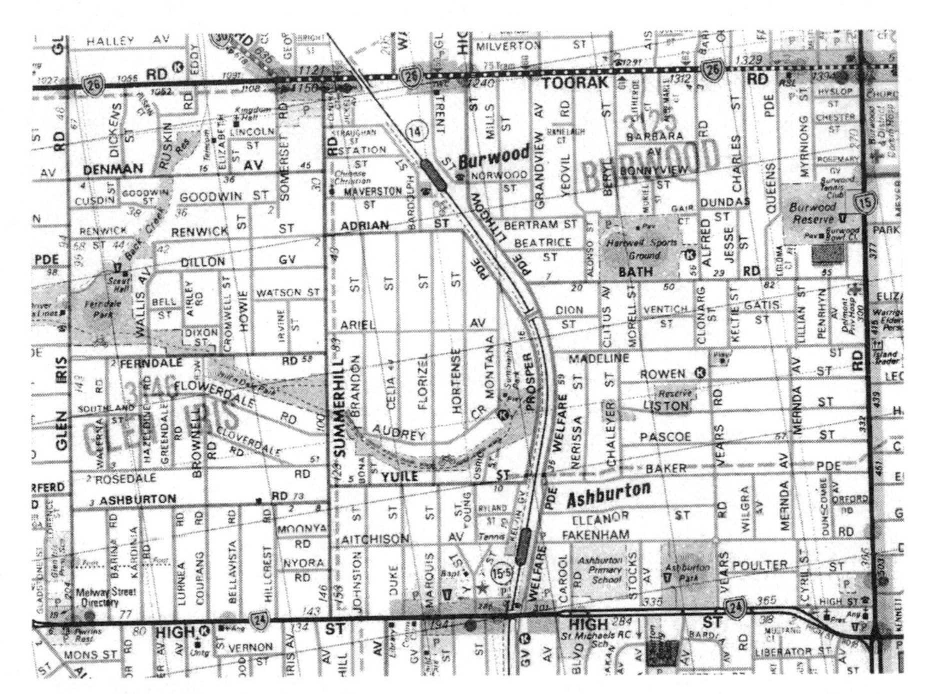

Figure 1: The Ashburton network.

Permeability Index

Consider two points P_0 and P on the main road sub-network H. Let $t(P_0, P)$ be the minimum travel time (or cost) from P_0 to P using the full network N, and let $T(P_0, P)$ be the minimum travel time (cost) from P_0 to P using sub-network H only. Then, by definition of a minimum path,

$$t(P_0, P) \leq T(P_o, P)$$

The dimensionless Permeability Index $\rho(P_0, P)$ may be defined as

$$\rho(P_o,P) = 1 - \frac{t(P_o,P)}{T(P_o,P)} \tag{3}$$

where $0 \leq \rho(P_0, P) \leq 1$. A value of zero means that G is impervious i.e. it offers no alternative cheaper path than H, while a value of one means that the local area sub-network G is completely pervious. This index offers a measure of the attractiveness of a local street system to through traffic. Comparing ρ values for different network designs offers a possible method for examining the relative effectiveness of the designs in limiting through traffic intrusion.

The index defined by equation (3) is a point-to-point measure. Performing a double integration over all links in the sub-network H to cover all possible origins and destinations in H yields an overall permeability index for the area,

$$\bar{\rho} = \frac{1}{L_H^2} \iint_{H\ H} \rho(P_o,P)dsdl \tag{4}$$

where L_H is the total road length in H. In practice, this integral is better replaced by a summation, which is easier to compute. Let $\bar{h}_i$ be the 'midpoint' of each link h_i in H. Similarly $\bar{h}_j$ is the 'midpoint' of each link h_j in H. A practical approximation to the overall permeability index for G can then be defined as

$$\bar{r} = \frac{1}{n_H^2} \sum_{ij} r_{ij} \tag{5}$$

where n_H is the total number of links in H and r_{ij} is given by

$$r_{ij} = 1 - \frac{t(\bar{h}_i,\bar{h}_j)}{T(\bar{h}_i,\bar{h}_j)} \tag{6}$$

The r_{ij} values may also be formed into a permeability matrix, which is also of interest in local area traffic planning. This matrix, for instance, can be used to assess differences in traffic movement in different directions across a study area network. A case study application of the permeability matrix is reported in this chapter.

Approachability Index

Traffic calming measures may be employed to reduce residential area network permeability and thus restrict through traffic intrusion into residential areas. Such measures may have undesirable side effects for local traffic movements in those networks. By definition, these traffic movements will involve trips with at least one trip end inside the residential area i.e. on sub-network G. Design of the traffic management scheme should focus on mitigating the adverse impacts on these 'essential' traffic movements whilst attempting to reduce the through traffic usage of the local street network. Local traffic splits into two main categories:

- traffic movements with one local trip end (on the sub-network G) and the other trip end outside the study area, i.e. on the sub-network H of main roads. For these trips the traffic planner may wish to examine the relative ease of access to or from the local area with a single crossing of its cordon line, and
- traffic movements that are entirely internal to the local area, i.e. that are made between two trip ends both located on the sub-network G. These movements will have quite different characteristics from the other traffic movements, principally that they do not need to use the main road sub-network H unless there are no acceptable alternative paths through sub-network G. Such situations may arise under traffic management schemes involving street closures or one-way street systems. In these cases frustratingly tortuous paths may be needed to complete the traffic movements.

Thus secondary indices considering the ability of the network to provide safe and efficient paths for local traffic movements are required, in addition to the Permeability Index described above. For traffic movements with one trip end outside the local street sub-network G, an Approachability Index may be defined to relate the spatial separation between the local trip end and the external trip end and the network-based distance between them. This index indicates the ease of accessibility from (to) a point P_0 on the main road network (H) to (from) a point Q on the local street network (G). The Approachability Index $\lambda(P_0, Q)$ is given by

$$\lambda(P_0, Q) = \frac{L(P_0, Q)}{S(P_0, Q)} \tag{7}$$

$$\lambda(Q, P_0) = \frac{L(Q, P_0)}{S(P_0, Q)} \tag{8}$$

where $L(P_0, Q)$ is the minimum spatial separation (e.g. Euclidean distance) between the points P_0 and Q, and $S(P_0, Q)$ is the path distance along the minimum network travel time (cost) between P_0 and Q. Path distances are more important for the Approachability Index as this is a parameter specific to the local traffic cell. It reflects the ease of movement to or from locations in the cell. The λ-index is bounded [0, 1] as $0 \leq L \leq S$. The larger the value of λ, the more direct (more approachable) is the path to the location. Note the existence of a reverse direction Approachability Index, from Q to P_0, as indicated by equation (8). It is quite possible that $\lambda(P_0, Q) \neq \lambda(Q, P_0)$.

Tortuosity Index

In the case of local traffic movements wholly within the local area, there is need for an index relating the spatial separation between the two local trip ends and the network-based distance (using network G) between them. A suitable index for purely internal travel in the traffic cell, analogous to the Approachability Index described above, is the Tortuosity Index $\sigma(Q_1, Q)$ which may be defined as[1]:

$$\sigma(Q_1, Q) = 1 - \frac{L(Q_1, Q)}{S(Q_1, Q)} \qquad (9)$$

where Q_1 and Q are both points in G and $L(Q_1, Q)$ and $S(Q_1, Q)$ are as defined for approachability. The Tortuosity Index is bounded [0, 1]. Higher values of σ indicate that the path is more tortuous, i.e. less direct. Unity as a value for the Tortuosity Index for two trip ends a finite distance apart would imply either an infinitely long network path, or a discontinuity in the network. A value of zero means that the minimum network path has the same path length as the Euclidean distance between the trip ends, i.e. it is a straight line.

The Approachability and Tortuosity Indices defined by equations (8) and (9) refer to pairs of points on the road network. As in the case of the Permeability Index, overall coefficients of approachability and tortuosity may be defined, as per equations (4) and (5) for both theoretical study and practical application.

Practical Approachability and Tortuosity Indices

The practical definitions of the Approachability Index and the Tortuosity Index are as follows.

---

[1] This definition of the Tortuosity index differs from the earlier definition proposed by Taylor (1999b). The new definition is based on semantic considerations, so that more tortuous (less direct) paths have higher values of the index.

Approachability. The practical value of the overall Approachability Index for a network is

$$\bar{l} = \frac{1}{n_H n_G} \sum_{ik} l_{ik} \tag{10}$$

where

$$l_{ik} = \frac{L(\bar{h}_i, \bar{g}_k)}{S(\bar{h}_i, \bar{g}_k)} \tag{11}$$

is an element of the 'approachability matrix' for the network and follows the definition of the approachability index given by equation (7). In equation (11) the $\bar{h}_i$ are the midpoints of the A-links in H and the $\bar{g}_k$ are the midpoints of the C-links in G.

Tortuosity. In analogous fashion the practical value for the overall Tortuosity Index for subnetwork G is defined as

$$\bar{s} = \frac{1}{n_G^2} \sum_{kl} s_{kl} \tag{12}$$

where the elements of the 'tortuosity matrix' are defined as

$$s_{kl} = 1 - \frac{L(\bar{g}_k, \bar{g}_l)}{S(\bar{g}_k, \bar{g}_l)} \tag{13}$$

using equation (9) and where $\bar{g}_k$ and $\bar{g}_l$ are the midpoints of the C-links and D-links respectively in G.

Case Study

The Ashburton area (Figure 1) in the eastern suburbs of Melbourne, Australia has been subject to a number of traffic management schemes over recent years. Residents' concerns about through traffic usage of the local street system were exacerbated in the late 1980s due to a nearby freeway construction project. The main feature of the residential area in Figure 1 is that it is bisected by a suburban railway line. There is only one crossing point (a bridge over the railway) inside the residential area network. The area is bounded by three major arterial roads: (1) Toorak Road along its northern perimeter, (2) Warrigal Road to the east, and (3) High Street to the south. The western edge is defined by Glen Iris Road, a secondary arterial road. A fourth arterial road, Highbury Road, terminates at Warrigal Road (see Figure 1). The new freeway passes to the west of the Ashburton area, with access from Toorak Road and from High Street. Previously the freeway terminated at Toorak Road, which is heavily congested for most of the day. The internal railway bridge offers a 'rat run', using Bath Road and Adrian Street to Summerhill Road, then splitting into two (Denman Street and Ferndale Avenue) for travel further west (Figure 1). The local council and residents were concerned that the new freeway interchange at High Street would draw commuter traffic to the south, to avoid the bottlenecks on Toorak Road, and thus create significant additional traffic intrusion inside the Ashburton area. A number of alternative traffic management plans were thus proposed to reduce the impact of through traffic on the local area. An evaluation of the alternative schemes using the TrafikPlan/MULATM model was reported by Taylor (1988).

TrafikPlan modelling of the alternative traffic management schemes allows the computation of permeability, approachability and tortuosity (PAT) indices for the area under different traffic conditions and network arrangements and using origin-destination movements assumed to represent travel demand in the area after the opening of the new freeway (Taylor, 1988). PAT analysis was undertaken of the following three schemes:

(1) The base case, being the existing (pre-1987) network, with little or no special traffic calming applied to the local street system;

(2) application of a traffic calming scheme designed by the council's traffic engineers to block the Bath Road-Adrian Avenue rat run. This scheme included the installation of speed control devices on some road links and partial closure of key local intersections, blocking some turns or even preventing some crossing manoeuvres, and

(3) the intuitively simple scheme of closing the internal railway bridge.

Closing the bridge is an effective way of severing the rat run, although it still leaves two segments of the run in place, and could cause significant disruption to local traffic movement given the distribution of land use facilities (e.g. shops and schools) in the area.

The PAT analysis was also made for two sets of traffic conditions: morning peak hour traffic and free flow traffic conditions. Travel time, delay and queuing data for the network were generated using TrafikPlan. Minimum travel time paths between selected origins and destinations could then be found and used to compute values for the PAT indices. These indices, as defined by equations (5), (10) and (12), would most properly be based on the computation of full permeability, approachability and tortuosity matrices covering the network. For purposes of brevity and illustration reduced matrices are reported here, each of size 5 × 5 (i.e. five origins and five destinations each). This chapter is intended only to illustrate the analysis method. The application of the PAT indices to the design of traffic calming plans for real networks would require a larger scale study. Nevertheless, some very interesting results emerge from the case study.

Thus five origins (A-links) and five destinations (B-links) on the arterial road network were selected for the computation of permeability indices. These are shown in Figure 2. Approachability indices were computed based on the five origins (A-links) on external arterial roads and five destinations (C-links) on local streets shown in Figure 3. Note that the A-links in Figure 3 are not the same as those in Figure 2. Tortuosity indices were based on the five origins (C-links) and five destinations (D-links) on the local street network in Figure 4. Computations were made for both the (congested) morning peak and for free flow traffic conditions, for the three alternative traffic management schemes.

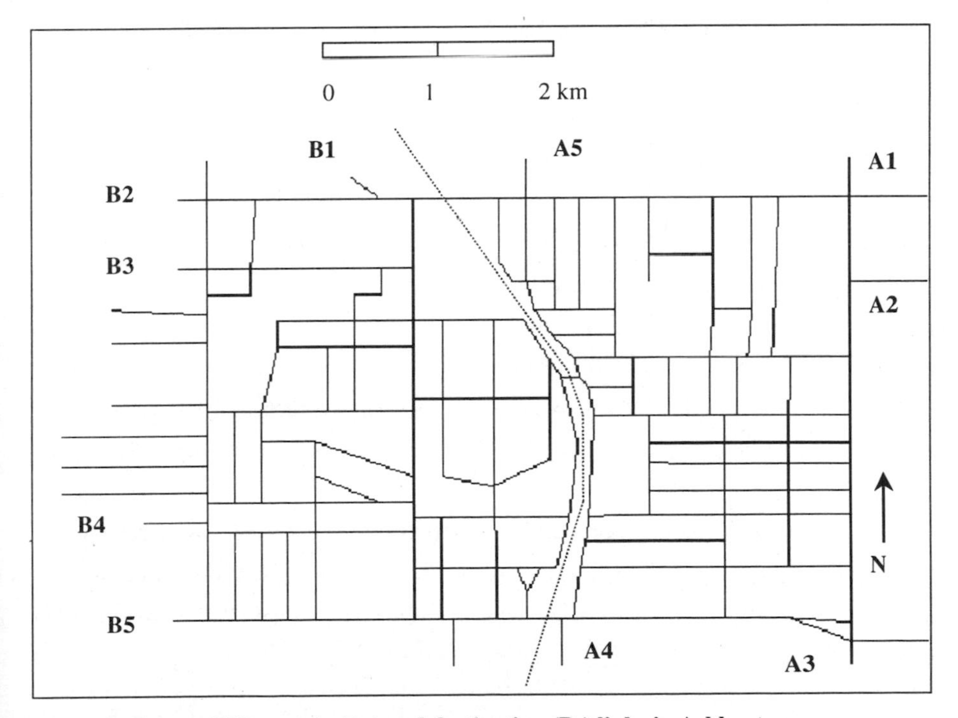

Figure 2: Permeability origin (A-) and destination (B-) links in Ashburton.

Permeability Indices

Table 2 shows the reduced permeability matrices for Ashburton in the morning peak. There are three separate matrices, one for each of the traffic management schemes. Mean values of the Permeability Indices are given, for each origin and each destination, as well as overall means. In the base case under morning peak traffic the overall mean Permeability Index for the area is 0.135. This reduces to 0.117 under the proposed traffic calming scheme, and to 0.103 with the bridge closure. Inspection of the origin and destination mean permeabilities show that not all of these trip ends show decreases in permeability despite the overall decrease. For example, origin A5 shows an increase in permeability. Here may be seen some of the power of the analysis method, which is able to indicate the relative effects of the different schemes on trip movements.

The permeability results for free flow traffic conditions are shown in Table 3. Under these conditions the area is impervious for the base case and for the bridge closure. A small level of permeability is seen for the scheme with traffic calming devices, for two trip movements (see the matrix in Table 3).

Approachability Indices

The permeability calculations suggest a tightening of traffic movement conditions under the two proposed traffic management alternatives to the base case. This is the desired result, but needs to be tempered by consideration of the effects on local traffic movements. Approachability is concerned with the relative ease of reaching specified destinations within the area, from external origins. Table 4 shows the reduced approachability matrices for the area under the different traffic management regimes in the morning peak. The mean Approachability Index for the base case is 0.614. This decreases to 0.571 under the proposed traffic calming scheme, and shows a slight rise (to 0.618) for the bridge closure. The higher the value of the Approachability Index, the 'easier' it is to reach a given internal destination. Under free flow conditions (see Table 5) the corresponding values of the overall mean approachability index are 0.606, 0.579 and 0.605.

Table 2: Permeability matrices for Ashburton under three alternative traffic calming plans, for the morning peak period.

Permeability matrix (ρ): Existing network, am peak

Origin (A-links)	Destination (B-links) 1	2	3	4	5	Mean ρ
1	0	0	0.041	0.149	0.158	0.070
2	0.172	0.166	0.240	0.275	0.162	0.203
3	0.265	0.188	0.093	0.021	0	0.113
4	0.240	0.136	0.036	0.032	0	0.089
5	0	0.161	0.211	0.323	0.316	0.202
Mean ρ	0.135	0.130	0.124	0.160	0.127	0.135

Permeability matrix (ρ): Council's traffic management scheme, am peak

Origin (A-links)	Destination (B-links) 1	2	3	4	5	Mean ρ
1	0	0.007	0.162	0.026	0	0.039
2	0.221	0.206	0.101	0	0	0.106
3	0.209	0.113	0.025	0	0	0.069
4	0.241	0.131	0.030	0	0	0.080
5	0	0.274	0.514	0.339	0.319	0.289
Mean ρ	0.134	0.146	0.166	0.073	0.064	0.117

Permeability matrix (ρ): Bridge closure, am peak

Origin (A-links)	Destination (B-links) 1	2	3	4	5	Mean ρ
1	0	0	0.033	0.033	0.123	0.038
2	0.143	0.131	0.156	0.134	0.139	0.141
3	0.250	0.123	0.023	0.012	0	0.082
4	0.230	0.069	0.006	0.007	0	0.062
5	0	0.225	0.256	0.235	0.237	0.197
Mean ρ	0.125	0.110	0.095	0.084	0.100	0.103

Table 3: Permeability matrices for Ashburton under three alternative traffic calming plans, for free flow traffic conditions.

Permeability matrix (ρ): Existing network, free flow travel time

Origin (A-links)	Destination (B-links) 1	2	3	4	5	Mean ρ
1	0	0	0	0	0	0
2	0	0	0	0	0	0
3	0	0	0	0	0	0
4	0	0	0	0	0	0
5	0	0	0	0	0	0
Mean ρ	0	0	0	0	0	0

Permeability matrix (ρ): Council's traffic management scheme, free flow travel time

Origin (A-links)	Destination (B-links) 1	2	3	4	5	Mean ρ
1	0	0	0.069	0	0	0014
2	0	0	0	0	0	0
3	0	0	0	0	0	0
4	0	0	0	0	0	0
5	0	0	0.494	0	0	0.099
Mean ρ	0	0	0.113	0	0	0.023

Permeability matrix (ρ): Bridge closure, free flow travel time

Origin (A-links)	Destination (B-links) 1	2	3	4	5	Mean ρ
1	0	0	0	0	0	0
2	0	0	0	0	0	0
3	0	0	0	0	0	0
4	0	0	0	0	0	0
5	0	0	0	0	0	0
Mean ρ	0	0	0	0	0	0

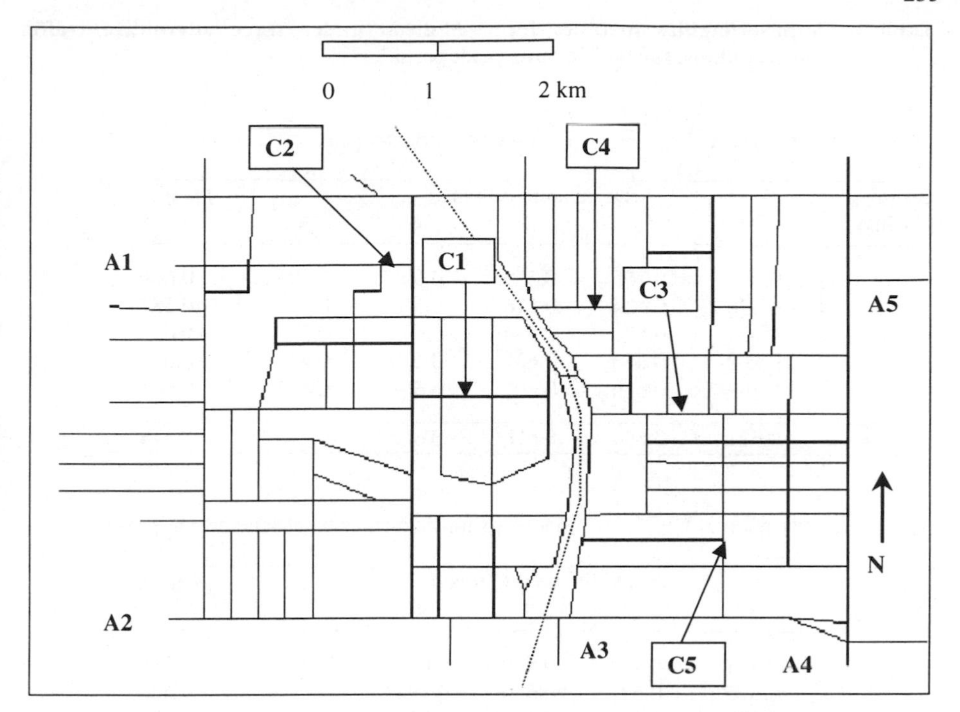

Figure 3: Approachability origin (A-) and destination (C-) links in Ashburton.

Thus it appears that the level of approachability in the network would be slightly reduced under the proposed traffic calming scheme, and little affected by the bridge closure. This may reflect the specification of the small number of C-links in the network? A full application of the method would need to be based on a much larger set of C-links.

Tortuosity Indices

Further investigation of changes in local travel may be made using the Tortuosity Index. In this case, following the definition of the index in equations (12) and (13), the higher the value of the index the more difficult (more tortuous) it is to reach the D-links. Figure 4 identifies the C-links and D-links used in the case study.

Table 4: Approachability matrices for Ashburton under three alternative traffic calming plans, for the morning peak period.

Approachability matrix (λ): Existing network, am peak

Origin (A-links)	Destination (C-links) 1	2	3	4	5	Mean λ
1	0.567	0.705	0.559	0.604	0.595	0.606
2	0.576	0.624	0.516	0.640	0.579	0.587
3	0.671	0.680	0.534	0.632	0.567	0.617
4	0.692	0.676	0.639	0.660	0.621	0.658
5	0.599	0.636	0.600	0.601	0.603	0.601
Mean λ	0.621	0.664	0.570	0.627	0.593	0.614

Approachability matrix (λ): Council's traffic management scheme, am peak

Origin (A-links)	Destination (C-links) 1	2	3	4	5	Mean λ
1	0.465	0.471	0.599	0.604	0.591	0.546
2	0.576	0.624	0.640	0.492	0.806	0.628
3	0.671	0.680	0.406	0.520	0:354	0.526
4	0.692	0.676	0.614	0.642	0.583	0.641
5	0.482	0.518	0.523	0.441	0.603	0.513
Mean λ	0.577	0.594	0.556	0.540	0.587	0.571

Approachability matrix (λ): Bridge closure, am peak

Origin (A-links)	Destination (C-links) 1	2	3	4	5	Mean λ
1	0.567	0.705	0.599	0.604	0.600	0.615
2	0.576	0.624	0.640	0.581	0.806	0.645
3	0.657	0.680	0.534	0.634	0.582	0.617
4	0.684	0.676	0.639	0.660	0.621	0.656
5	0.400	0.567	0.600	0.601	0.605	0.555
Mean λ	0.577	0.650	0.602	0.616	0.643	0.618

Table 5: Approachability matrices for Ashburton under three alternative traffic calming plans, for free flow traffic conditions.

Approachability matrix (λ): Existing network, free flow travel time

Origin (A-links)	Destination (C-links) 1	2	3	4	5	Mean λ
1	0.567	0.705	0.559	0.598	0.600	0.614
2	0.576	0.624	0.640	0.489	0.806	0.627
3	0.657	0.680	0.534	0.634	0.582	0.617
4	0.684	0.574	0.614	0.483	0.621	0.595
5	0.463	0.684	0.600	0.521	0.603	0.574
Mean λ	0.589	0.653	0.597	0.545	0.642	0.606

Approachability matrix (λ): Council's traffic management scheme, free flow travel time

Origin (A-links)	Destination (C-links) 1	2	3	4	5	Mean λ
1	0.465	0.471	0.599	0.598	0.466	0.520
2	0.576	0.513	0.640	0.489	0.806	0.605
3	0.657	0.680	0.534	0.632	0.582	0.617
4	0.684	0.574	0.614	0.483	0.621	0.595
5	0.463	0.684	0.523	0.521	0.603	0.559
Mean λ	0.569	0.584	0.582	0.545	0.616	0.579

Approachability matrix (λ): Bridge closure, free flow travel time

Origin (A-links)	Destination (C-links) 1	2	3	4	5	Mean λ
1	0.567	0.705	0.599	0.598	0.600	0.614
2	0.576	0.624	0.640	0.489	0.806	0.627
3	0.657	0.680	0.534	0.632	0.582	0.617
4	0.684	0.574	0.614	0.483	0.621	0.595
5	0.463	0.684	0.600	0.521	0.603	0.574
Mean λ	0.589	0.650	0.597	0.545	0.642	0.605

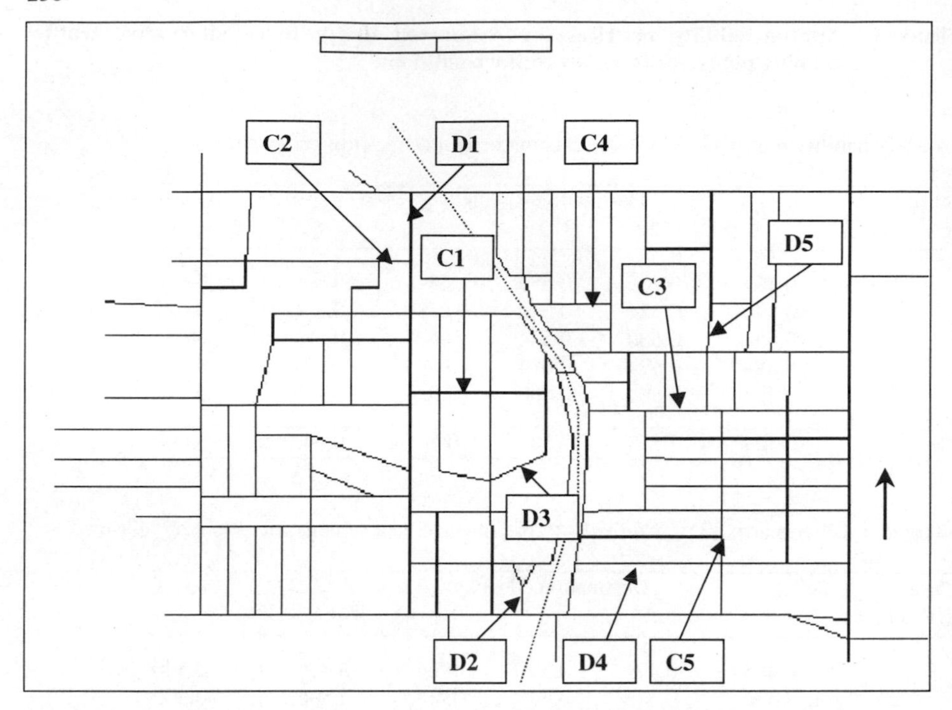

Figure 4: Tortuosity origin (C-) and destination (D-) links in Ashburton.

Table 6 indicates the computed indices for the morning peak period. The overall mean Tortuosity Index for the base case is 0.426. This increases to 0.436 under the proposed traffic calming scheme, and to 0.454 with the bridge closure. The effects of the bridge closure on particular origins and destinations (e.g. D3 and A4) may be seen in the table. Table 7 shows the indices for free flow traffic conditions. The overall index values for free flow are 0.416 for the base case, 0.419 for the traffic calming scheme, and 0.459 for the bridge closure. Under free flow the network is thus slightly less tortuous, except for the bridge closure. Note that the selection of D-links is based on the locations of significant local facilities, such as shopping centre car parks (D1 and D2), kindergartens (D3 and D5) and the local primary school (D4). Bridge closure might well have significant impacts on families living in the area.

Table 6: Tortuosity matrices for Ashburton under three alternative traffic calming plans, for the morning peak period.

Tortuosity matrix (σ): Existing network, am peak

Origin (C-links)	Destination (D-links) 1	2	3	4	5	Mean σ
1	0.399	0.149	0.551	0.652	0.384	0.427
2	0.682	0.207	0.408	0.543	0.445	0.457
3	0.370	0.486	0.473	0.391	0.421	0.428
4	0.635	0.387	0.476	0.564	0.433	0.499
5	0.388	0.328	0.601	0.142	0.135	0.319
Mean σ	0.495	0.311	0.502	0.458	0.364	0.426

Tortuosity matrix (σ): Council's traffic management scheme, am peak

Origin (C-links)	Destination (D-links) 1	2	3	4	5	Mean σ
1	0.399	0.149	0.551	0.652	0.684	0.487
2	0.682	0.207	0.408	0.543	0.445	0.457
3	0.369	0.460	0.471	0.391	0.421	0.422
4	0.635	0.358	0.476	0.564	0.433	0.493
5	0.400	0.328	0.611	0.142	0.135	0.323
Mean σ	0.497	0.300	0.503	0.458	0.423	0.436

Tortuosity matrix (σ): Bridge closure, am peak

Origin (C-links)	Destination (D-links) 1	2	3	4	5	Mean σ
1	0.399	0.149	0.551	0.652	0.690	0.488
2	0.682	0.203	0.408	0.543	0.445	0.456
3	0.366	0.421	0.731	0.391	0.421	0.466
4	0.425	0.440	0.763	0.564	0.433	0.525
5	0.385	0.398	0.610	0.142	0.135	0.333
Mean σ	0.451	0.322	0.613	0.458	0.425	0.454

Table 7: Tortuosity matrices for Ashburton under three alternative traffic calming plans, for free flow traffic conditions.

Tortuosity matrix (σ): Existing network, free flow travel time

Origin (C-links)	Destination (D-links) 1	2	3	4	5	Mean σ
1	0.399	0.149	0.551	0.589	0.369	0.411
2	0.682	0.207	0.408	0.514	0.468	0.456
3	0.377	0.471	0.473	0.391	0.421	0.427
4	0.391	0.358	0.476	0.564	0.433	0.444
5	0.530	0.328	0.583	0.142	0.135	0.344
Mean σ	0.476	0.303	0.498	0.440	0.365	0.416

Tortuosity matrix (σ): Council's traffic management scheme, free flow travel time

Origin (C-links)	Destination (D-links) 1	2	3	4	5	Mean σ
1	0.399	0.149	0.551	0.589	0.384	0.414
2	0.682	0.207	0.408	0.514	0.468	0.456
3	0.426	0.471	0.473	0.391	0.421	0.436
4	0.391	0.358	0.476	0.564	0.433	0.444
5	0.530	0.328	0.583	0.142	0.135	0.344
Mean σ	0.486	0.303	0.498	0.440	0.368	0.419

Tortuosity matrix (σ): Bridge closure, free flow travel time

Origin (C-links)	Destination (D-links) 1	2	3	4	5	Mean σ
1	0.399	0.149	0.551	0.688	0.700	0.498
2	0.682	0.207	0.408	0.514	0.468	0.456
3	0.377	0.471	0.473	0.391	0.421	0.481
4	0.391	0.440	0.758	0.564	0.433	0.517
5	0.530	0.328	0.583	0.142	0.135	0.344
Mean σ	0.476	0.319	0.609	0.460	0.431	0.459

The reduced matrices used in the computations for this case study are for illustrative purposes only. Larger matrices would be needed for a full analysis of the area and the implications of the traffic management alternatives. And a full analysis would also need to consider traffic conditions at other times of day, including the evening peak. The analysis presented here indicates that the PAT analysis has potential as an analytical tool to assist in the appraisal of traffic impacts of alternative network designs and traffic management plans.

DISCUSSION AND CONCLUSIONS

Concepts of network travel time reliability form the basis for the development of a set of network reliability indices that may be used to test and compare alternative network plans. These indices have the potential for use in multicriteria assessment where traffic management measures may be sought to limit through traffic usage of local streets whilst providing an adequate level of accessibility for local traffic. There is considerable scope for further research to examine, apply and extend the indices. An obvious application for this approach to network reliability assessment is in the planning and design of local street networks and the comparisons between different network topologies and traffic management plans, as well as different traffic conditions and travel patterns.

The case study presented in this chapter gives indications that the PAT analysis procedure has promise as a tool for the appraisal of alternative traffic management plans and network configurations. The indices appear capable of highlighting changes in network performance and in differentiating between effects on through traffic and local traffic. Further exploration of the procedure is warranted, to examine the capacity to compute the full PAT matrices for a given network, and the potential to use reduced matrices (of smaller dimension) – as in the case study – to analyse changes in performance.

Another area of potential application is in incident planning for congested networks. The PAT indices could form part of a toolkit for undertaking ‘congestion audits’, by predicting the relative impacts on network performance of blockages or bottlenecks caused by incidents occurring at different locations within the network. This audit might be used to identify congestion 'soft spots', where blockages could have more severe impacts than at other locations.

REFERENCES

Bell, M G H and Iida, Y (1997). *Transportation Network Analysis*. John Wiley and Sons, Chichester.

Brindle, R E (1989). SOD the distributor! *Multi-Disciplinary Engineering Transactions of the Institution of Engineers, Australia* **13** (2), 99-112.

Brindle, R E (1991). Traffic calming in Australia: a definition and commentary. *Australian Road Research* **21** (2), 37-55.

Eliot Hurst, M E (1974). *Transportation Geography: Comments and Readings*. McGraw-Hill, New York.

Herman, R E and Lam, T (1974). Trip characteristics of journeys to and from work. In: *Transportation and Traffic Theory* (D J Buckley, ed), pp.57-85. AH and A W Reed, Sydney.

Marshall, S (1998). Towards the integration of urban transport networks and urban design. *Proc 8th World Conference on Transport Research*. Antwerp, July.

Pfefer, R C and Raub, R A (1998). Effects of urban roadway blockage on capacity. *ITE Journal* **68** (1), 46-51.

Richardson, A J and Taylor, M A P (1978). A study of travel time variability on commuter journeys. *High Speed Ground Transportation Journal* **12** (1), 77-99.

Taylor, M A P (1979). Evaluating the performance of a simulation model. *Transportation Research A* **13A** (3), 159-173.

Taylor, M A P (1982). Travel time variability – the case of two public modes. *Transportation Science* **16** (4), 517-521.

Taylor, M A P (1988). MULATM and the SEMARL project. II: model analysis and application. *Traffic Engineering and Control* **29** (3), 135-141.

Taylor, M A P (1992). *User Manual for TrafikPlan Version 4.0*. School of Civil Engineering, University of South Australia, Adelaide, Australia.

Taylor, M A P (1999a). Dense network traffic models, travel time reliability and traffic management. I: general introduction. *Journal of Advanced Transportation* **33** (2), 218-233

Taylor, M A P (1999b). Dense network traffic models, travel time reliability and traffic management. II: application to network reliability. *Journal of Advanced Transportation* **33** (2), 235-251.

Westerman, H L (1990). Roads and environments. *Australian Road Research* **20** (4), 5-23.

Wong, H K and Sussman, J M (1973). Dynamic travel time estimation on highway networks. *Transportation Research* **7**, 353-370.

Keyword Index